First Steps in Rad

by Doug DeMaw, W1FB

Published by the
AMERICAN RADIO RELAY LEAGUE
225 Main Street
Newington, CT 06111

Printed in USA

ISBN: 0-87259-228-6

First Edition
Third Printing, 1993

Foreword

Amateur Radio. It is a rewarding and enjoyable hobby that enables ordinary people around the world to communicate at the touch of a button. With today's electronic technology, nothing is impossible! Take your pick—operate through amateur satellites or even by bouncing signals off the moon; send a lengthy letter to a friend by computer over packet radio; enjoy a leisurely voice or Morse code chat with an amateur across town or across the ocean, or even exchange pictures. Underlying all of these technological marvels are the basic concepts of electronics and radio theory. The purpose of this book is to open the doors to those who wish to learn more about the technical side of Amateur Radio.

Originally appearing in 1984 and 1985 issues of *QST*, the wide-ranging First Steps in Radio series helped newcomers to learn the electronic theory needed for licensing exams and to gain some insight into how their radio equipment works. The entire *QST* series is reproduced here. In the following pages, you will find basic explanations of circuit components, see these components assembled into practical circuits, and see how the circuits make up your radio gear. Additional segments cover antennas, propagation and radio-frequency interference at a beginner's level.

Author Doug DeMaw, W1FB, is a familiar name to many. On the ARRL technical staff for 18 years, Doug has authored more than 200 articles. Research in various areas has earned him patents. Doug's favorite Amateur Radio pastime is QRP (low power) operation, and to the delight of his readers he has designed many pieces of easy-to-build QRP equipment.

We hope you will learn about the technical side of Amateur Radio as you turn the pages of this publication. Although you may never "roll your own" equipment, you will gain an understanding of what goes on behind the front panel and take pride in that knowledge.

David Sumner, K1ZZ
Executive Vice President

Newington, Connecticut
October 1985

Contents

Getting into Amateur Radio Electronics

Part 1: Ever wonder what you need to know to pass your first amateur exam? This book will provide the answers — not to the FCC questions themselves, but to the questions most newcomers have about electronics.

Let's face it: Many potential amateurs feel a bit wary of tackling the electronics involved with earning that first ticket. Whether you're a housewife, a janitor, a factory worker, an English teacher or an advertising executive you may feel inadequate when the time comes to study for an amateur exam. That feeling seems to be shared by most people without a formal background in electronics, no matter where in the world they may live.

After reading the articles in this series, you'll find out for yourself that anyone with the motivation to learn the electronic theory needed for an Amateur Radio license can do so — regardless of their background. I've known children under 10 years of age who passed the Novice exam on the first try, and I've been acquainted with amateurs who were over 80 when they obtained their first license. And then there are persons with disabilities — those without sight or hearing (or both) who have progressed from the first license to the highest class of license (see sidebar on the different types of amateur licenses). Certainly, they have traveled a route that was far more rocky than those of us with no physical impairments.

A great many aspirants seem to give up before they give it a fighting chance. Others attempt to memorize the answers to exam questions. This practice has worked for some people, but it is not to their long-term advantage. Understanding the fundamentals — and that's what it amounts to — of Amateur Radio electronics is very important if you are to feel confident at exam time. This basic knowledge will prove invaluable later in your ham career, too: You'll be able to service your own equipment, you won't be afraid to discuss circuits at club meetings and on the air, and you can enjoy one of the special thrills of ham radio by experimenting and building some of your own equipment.

We shouldn't ignore still another benefit of knowing Amateur Radio theory: It's been the stepping stone to a career in electronics for countless young people. Furthermore, possession of a license puts you in a position to be of service to the federal, state and community governments in time of emergency or disaster. You can be a valuable resource in time of need.

The Fundamentals of Electricity

You may have studied basic electrical theory in high school, but you may have forgotten it because it didn't pertain to your present way of life. That happens to a great many people. So, let's discuss some very fundamental concepts. We'll get into a more detailed treatment in future installments of this series. But for the present, let's talk about ac and dc voltages and currents. These are the basis of all electronics theory, so they are mighty important to us.

Voltage means potential difference. It is called *potential* because the electrical charge is capable of doing some work but

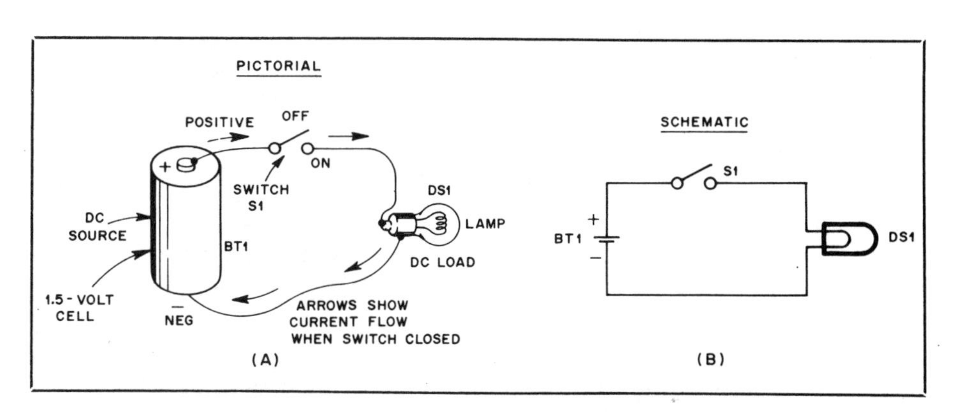

Fig. 1 — The illustration at A shows a simple dc circuit in pictorial form. The arrows indicate the direction of current flow. The drawing at B is the same circuit but presented in schematic form.

Glossary of Terms

ac — alternating current, or electrical current that flows in one direction, then in another.
ampere — the unit of electrical current, abbreviated A.
ARRL — The American Radio Relay League, Inc., headquarters for U.S. and Canadian ham radio operators and the society of the International Amateur Radio Union.
current — the flow of electrons.
CW — continuous wave, or Morse code.
dc — direct current, or electrical current that flows in only one direction.
Hz — the abbreviation for hertz, one cycle per second.
IEEE — The Institute of Electrical and Electronics Engineers, a professional society.
kHz — the abbreviation for kilohertz, 1000 hertz.
MHz — the abbreviation for megahertz — 1 million hertz.
oscilloscope — a device for giving a visual trace of voltage with respect to time; often called a scope, for short.
QSO — contact with another radio amateur.
QST — the official journal of The American Radio Relay League; also a general call preceding a message addressed to all amateurs and ARRL members.
RF — radio frequency.
transformer — a device for converting voltage levels.
voltage — electrical pressure causing electron flow.
volt — the unit of voltage, abbreviated V.
watt — the unit of power, abbreviated W.

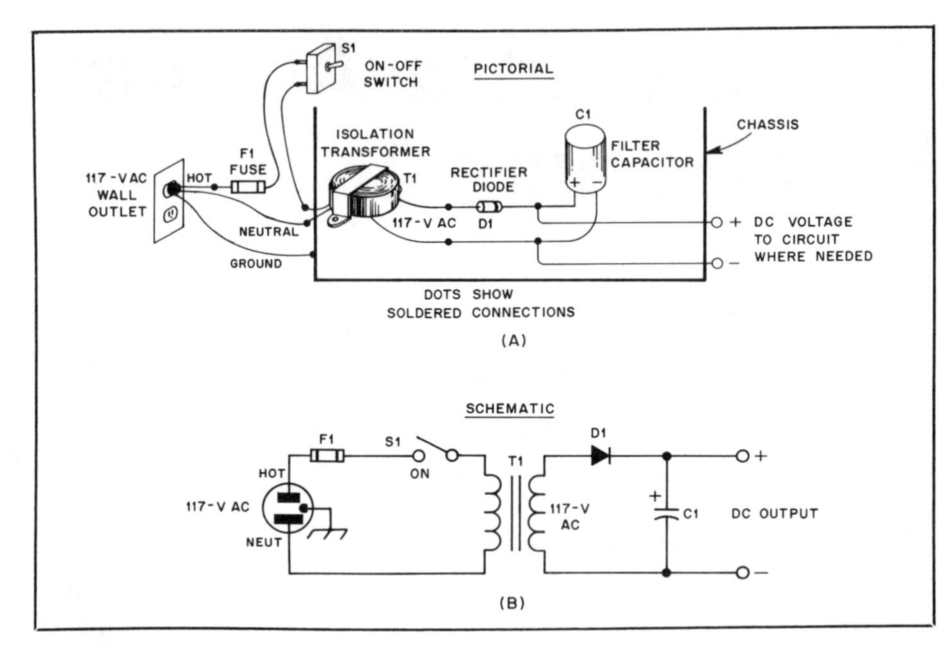

Fig. 2 — A pictorial diagram (A) of a dc power supply that is operated from the standard wall outlet (117-V ac). D1 changes the ac voltage to pulsating dc voltage, and capacitor C1 removes the small amount of ripple that remains after rectification. The same circuit is shown at B in schematic form.

may or may not be doing work. Voltage is also called "electromagnetic force." That may be a mouthful, but the idea is that voltage is an electrical pressure or force ready to be put to work.

Current is flow of electrons. Electron flow can take place only when there is a voltage (potential difference) and a conductor through which to move. As an analogy, picture two adjacent lakes that we'll call High and Low. Lake High has a water level that is several feet higher than Lake Low. If we cut a small channel between them but put a lock in the channel, no water will flow. But there will be a pressure difference. When we open the lock, water will flow from Lake High to Lake Low until the difference is gone and both lakes are the same level. In electricity, water level in this analogy is similar to voltage, and current is similar to water flow. Electrical current will flow until the potential difference is eliminated or the path is blocked.

DC (direct current) is defined as "A unidirectional current in which the changes in value are either zero or so small that they may be neglected.[1]

What this means is that if we could see dc with our eyes, it would flow in only one direction (like a river) and would appear as a straight line with no humps or bumps. We can see this if we hook up an oscilloscope to the dc-voltage line. The dc will show up on the face of the scope tube as a straight line. Fig. 1 provides a simple illustration of direct current and how it flows. Common sources of dc voltage are flashlight and car batteries. Only dc can be stored in batteries.

We can change alternating currents, the kind of electricity used in homes and business, to dc voltage by *rectifying* and *filtering* it. For example, we can take the voltage from a standard wall outlet (117-V ac), connect it to a transformer (a safety measure to protect us from the high-current voltage source), then pass the ac through a tube or semiconductor rectifier diode. This will give us *pulsating dc* voltage because some of the ac will still be present. These remaining small pulses can then be removed almost entirely by adding a filter capacitor after the rectifier. A simple example of this is given in Fig. 2. A transformer can also be used to increase (step up) the ac-line voltage or lower it (step down).

AC (alternating current) is defined as, "A periodic current the average value of which over a period is zero. *Note:* Unless distinctly specified otherwise, the term *alternating current* refers to a current that reverses at regularly recurring intervals of time and that has alternately positive and negative values."

What does all of this jargon mean? Simply, that ac voltage has a starting point (zero reference) of no value (zero voltage). Then it rises to a particular peak (high) value, falls back to zero and then drops to a negative value that is equal to the peak value. This rise and fall occurs at precise periods. For example, the voltage from our ac wall outlets is rated at 50 or 60 Hz (hertz), also called "cycles per second." This means the current will travel through one complete cycle — zero to plus, plus to zero, zero to negative and negative back to zero — in a given length of time. This happens 60 times per second with the current from our wall outlet, and may occur several million times per second with the radio-frequency energy that amateurs use to communicate. An ac cycle is illustrated in Fig. 3.

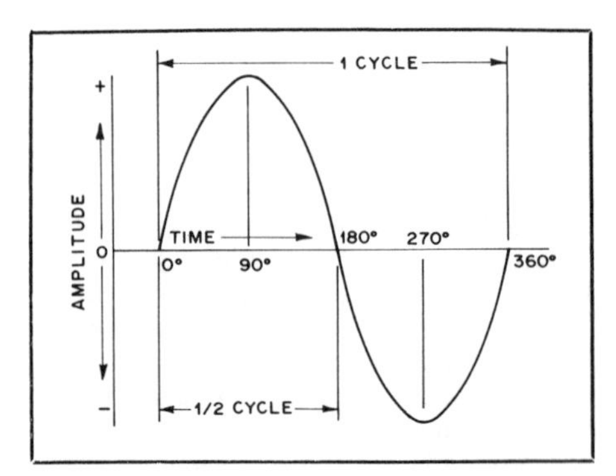

Fig. 3 — Representation of ac voltage, showing how it commences at zero, swings positive, returns to zero, swings negative and returns again to zero. This represents one complete ac cycle.

Ac is used mainly to power our homes, to illuminate the bulbs in our lamp, and to operate motors, stoves and the like. On the other hand, most electronic equipment requires a dc-voltage source. So, we feed the ac to a *power supply* (Fig. 2), which changes it to direct current.

The power lines that feed our homes, and that we see crossing the highways and countrysides, carry ac voltage. Some of them convey thousands of volts from the

U.S. Amateur License Classes

There are five classes of Amateur Radio license in the U.S.: Novice, Technician, General, Advanced and Extra. Although most beginners start with the Technician, you can start with any license class. Let's take a look at the requirements for and privileges of each class of license.

Novice: To earn a Novice ticket, you'll need to pass a 30-question written test and a 5-word-per-minute Morse code test. Novices can use Morse code on four HF (high-frequency) bands. With a Novice license, you can also operate voice and digital modes like packet radio on one of these bands. The HF bands can allow long-distance (or even worldwide) communication—when atmospheric conditions and your station equipment are right. Novices can also use repeaters and other communications modes on the 222- and 1270-MHz bands.

Technician: The beginner's license of choice, the Technician is the only license that does not require a knowledge of the Morse code. To earn this license, you need to pass the Novice written (30 questions) plus a 25-question Technician test. With this license, you can operate on all the VHF, UHF and microwave bands hams can use. The reason this license is so popular is that it allows operation on 2-meter repeaters (used for local communication) and packet radio (a mode that uses your computer to communicate over a network, either locally or over long distances). Technicians who pass a 5-word-per-minute Morse code test also have Novice HF privileges.

General: A General class license gets you on all of the HF bands hams can use, including the coveted 20-meter band. The written exam consists of 25 questions, and the Morse code test is at 13 words per minute. If you start with the General class license, you'll need to pass the Novice and Technician written exams, too.

Advanced: To qualify for this license you need to pass an additional 50-question written exam and (if you haven't already passed it) a 13-word-per-minute code test. Advanced-class licensees can use larger portions of the HF bands than Generals.

Extra: This license gives you all of the many operating privileges available to hams. There is an additional 40-question written test as well as a 20-word-per-minute Morse code test.

For more information on license requirements, study materials and exam opportunities, contact ARRL Educational Activities Department, 225 Main St, Newington CT 06111-1494, tel 203-666-1541. Prospective amateurs call 1-800-32-NEW HAM (1-800-326-3942).

generating plants to communities many miles distant. This high ac voltage is lowered before it enters our homes. A step-down transformer (located on a nearby power pole) is used for this purpose. The principle of operation for the "pole transformer" is identical to that of the transformer in a dc power supply. The notable difference is in the high amount of power the pole transformer can accommodate. Also, we do not rectify the output from the pole transformer to turn it into dc.

As we mentioned earlier, the RF (radio frequency) energy that amateurs feed to their antennas when transmitting is also ac, but the cyclic rate is very high. For example, a 3500-kHz radio signal goes through its ac cycle 3.5 million times a second. Audio energy (sound waves) is also ac, and the cyclic rate varies constantly when the human voice (or music) is reproduced. The frequency depends on the particular tone at a given instant.

The Matter of Power

Thus far we have discussed voltage and current. But, what about power? In broad terms we tend to think about "power" as a reserve of strength we may call on to perform a task. Car engines are rated in terms of power, or horsepower. Or, someone might say, "He is a powerful man." In the electrical world, power is "the rate of doing work." It is equal to the voltage multiplied by the current. This relationship can be expressed as a simple equation: $P = E \times I$, where P is the power in watts, E is the voltage in volts and I is the current in amperes. Thus, if we had a light bulb that operated from 120 volts, and it required a current of 0.83 ampere to illuminate fully, the bulb would consume 100 watts of power when lit.

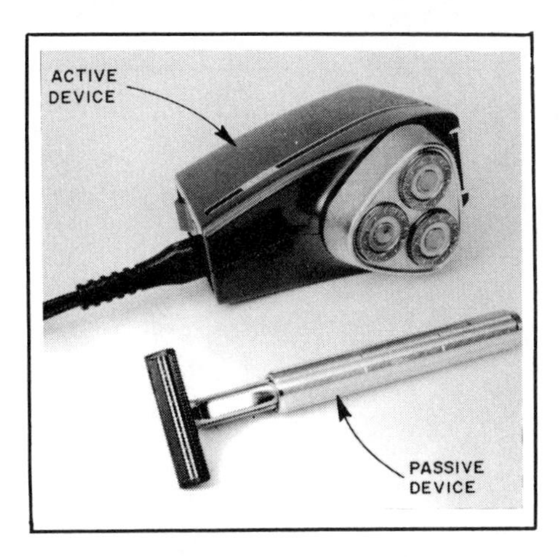

Fig. 4 — Simple pictorial illustration of an active and a passive device (see text).

We can see from this that the higher the power consumption of a circuit or appliance, the greater the available current requirement. Power, current and voltage are, therefore, the basis of all electrical circuits. The notable exception is when we use what is called a *passive* circuit, one that requires no operating voltage (and therefore does not consume power). Such circuits do have a maximum voltage, current or power rating, though. This means that we dare apply only a certain amount of signal energy to them, lest they be destroyed by excessive power dissipation, caused by current flowing through them. A circuit or device that requires an operating voltage (and draws current) is called an *active* circuit. (See Fig. 4.)

Getting it Together

If you stayed with me through this discussion, you should have a better understanding of the basics of electricity.

At this juncture you may be saying to yourself, "Sure, it's easy for him to say how easy it is. After all, he's been in this game for a long time!" Well, let me tell you how *I* got started. I was an 8th-grade student when two other fellows and I happened across a book in the school library that described early-day transmitters. We built homemade spark-gap transmitters and antennas from that book, then went blithely on the air, not realizing that a license was required!

Later in life, after getting over the trauma caused by my experience as a

"bootlegger" illegal operator, my interest in radio was rekindled after watching the shipboard operators during WW II. I knew no hams and had no background in electronics. I obtained a copy of *QST,* then borrowed an old *ARRL Handbook.* I was off and running! A friend let me borrow her Webcor disk recorder, which I used to transcribe my own CW sending (after I learned the code with a hand key. I recorded some pages from *QST,* but put the text on the disks backward, starting at the bottom of the page and working toward the top. This prevented me from memorizing the text. Meanwhile, I sent for an ARRL *License Manual,* and between that and the *Handbook* I prepared for the amateur exam. A month later, I went to the Detroit FCC office and passed my test to become

WN8HHS. I met my first ham on the air! So, I know from experience that if one really wants to be a ham, it can be done — whether or not that person has a knowledge of electrical circuits and FCC regulations.

I hope you've been inspired toward taking that first step into the world of Amateur Radio. Read on for more basic theory and its practical application to Amateur Radio.

Note

[1]*IEEE Standard Dictionary of Electrical and Electronic Terms,* published by Wiley-Interscience, a division of John Wiley and Sons, Inc., New York, NY.

How to Read a Schematic Diagram

Part 2: The first step toward learning the basic theory of this series is to understand circuit diagrams — the "road maps" that

allow us to build or repair equipment.

"Sure, I can handle the electronics — up to a point. I start having problems when I try to figure out what's going on in schematic diagrams." Many newcomers to radio electronics have this problem. Perhaps you're one of them.

In this installment, we'll learn what the various electronics symbols stand for, and we'll get a feel for how the diagram relates to the actual circuit-board layout.

Learning the Symbols

We must first accept the fact that very few electronics symbols look like the physical item they represent. Only a *pictorial* diagram can satisfy that requirement. Most electronic parts are encased or encapsulated in some manner, which prevents us from peering inside to see what is there. Semiconductors (diodes, transistors and integrated circuits) are the worst in this regard, for if we did saw one open for a look-see, we might be hard-pressed to recognize the various elements (drain, base, emitter, collector, source, gate, or whatever) unless we understood the philosophy of semiconductor design and fabrication. So, our best approach is to ignore for now the contents of the enclosed components and think mainly about how the leads relate to the inner elements, as defined by the assigned symbol. In the days of vacuum tubes we could dismantle a tube and easily identify the grid, plate, cathode and filaments, but things have changed!

Unfortunately, each publisher of amateur and commercial electronics magazines or journals follows his or her own symbology. For this reason, diagrams found around the world can conflict. The ARRL has adopted and used the IEEE (Institute of Electrical and Electronic Engineers) standard symbols for many years. Only a few exceptions exist, and that is done to simplify (or unclutter) the drawings in *QST* and other League publications. We will focus on the *QST* symbology here, and despite differences found in other publications, you should be able to determine what a symbol stands for, because there will be ample similarity. Some magazine publishers, in order to establish a distinctive "style," have more or less ignored the recommended standards for electrical symbols. It is unfortunate, but we must accept it.

An abbreviated presentation of electrical symbols is provided in Fig. 1. You will see that some symbols do, indeed, resemble what they stand for, such as the headset, speaker and hand key. Conversely, the symbols for ICs (integrated circuits) would in some instances fill one or two *QST* pages if we were to see all of what was inside the IC. So for these complex circuits we accept the practical solution — to use just a box, a triangle or similar representation. In a real-life situation we think only about where each external lead connects, according to the numbers assigned to the pins by the manufacturer. This was for many years known as the "black box" approach. In other words, don't worry about what's within the box; just concentrate on what the box will do for us.

You will note that for some symbols we have more than one format. This means that we may use any of the illustrations given, and we may find one or all of them in a single issue of *QST*. The wiring junctions at the lower right of Fig. 1 are an example of what we are discussing.

The best advice I can offer at this time is to spend a few evenings studying and memorizing the symbols in Fig. 1. When you feel that you have the data implanted

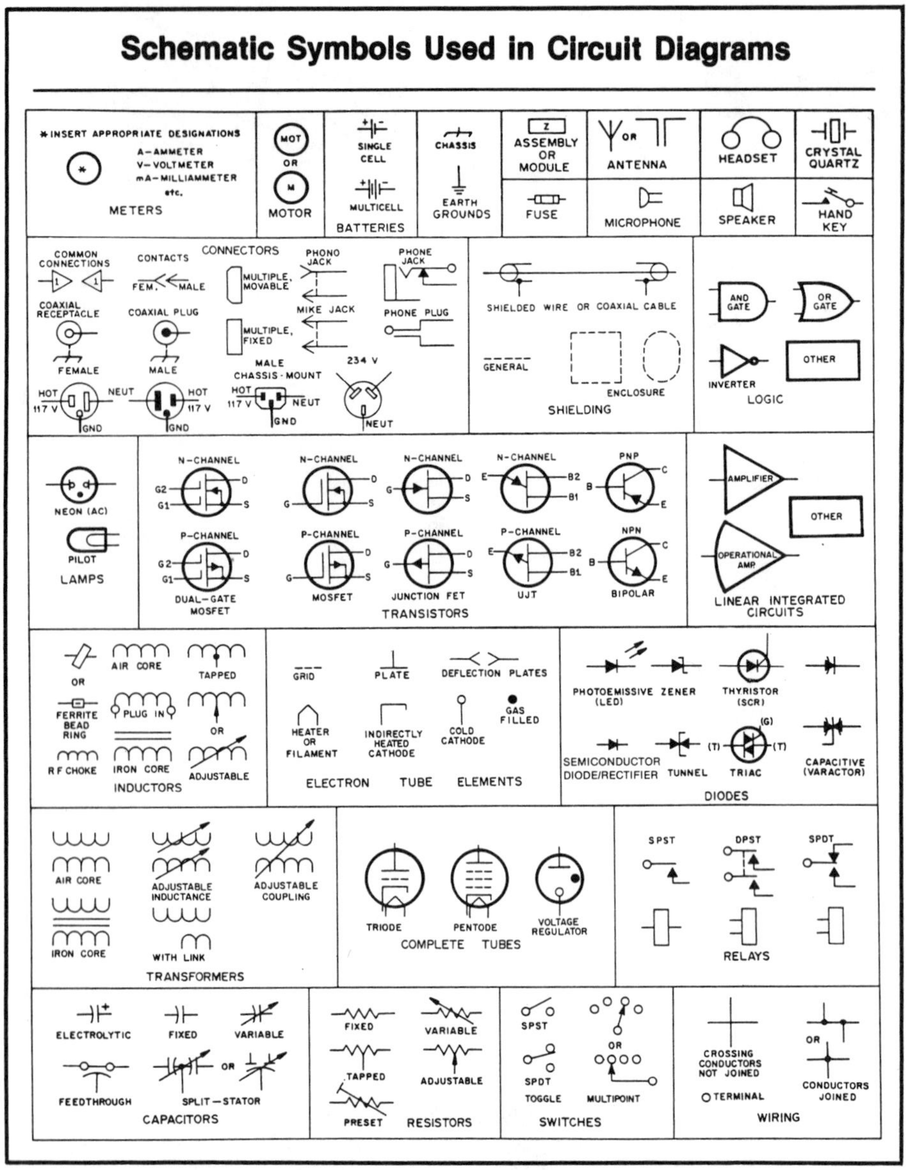

Schematic Symbols Used in Circuit Diagrams

Fig. 1 — Collection of standard symbols used by the ARRL for circuit diagrams. Most of these symbols were adopted from the IEEE standards.

firmly in your mind, put away the symbols page and try to draw each symbol by memory, writing its name next to it. Continue with the exercise until you make no mistakes. This knowledge will prove invaluable to you as you pursue that ham license. It will be helpful to know the symbols after you pass the exam, also. You will need to have this knowledge in order to repair your equipment, to duplicate home-construction projects in the magazines, or to do your own circuit designs. If you are a person without sight, have a friend provide you with word pictures of the symbols and learn them that way. I know at least two blind amateurs who repair their own equipment by having someone give them word pictures of the diagram section that applies to the problem.

A Simple Circuit Example

Let's try our luck at relating a simple circuit to a pictorial diagram. This will enable you to see how things hook together when assembling a circuit from a schematic diagram. Fig. 2 shows two schematic diagrams of a two-stage audio amplifier, such as we might find in the early stages of a receiver. Although the two drawings look quite different at first perusal, you will observe that they represent the same circuit in complete detail. The difference is only in the manner of illustrating the circuit schematically. Fig. 2A shows the ground connections separately. Likewise with the two +12-volt connections. Fig. 2B shows the ground and +12-volt lines joined together, respectively. The net result is the same in either case: In a practical circuit the ground or +voltage lines would eventually be joined at a common point when example A is followed. It merely illustrates that you may find more than one style of presenting a circuit. You will note also that the resistors which connect to the +12-volt source (Fig. 2B) are routed upward rather than downward, as in Fig. 2A. You may find a mixture of the two methods in a given drawing, so don't let that confuse you. The main objective is to make sure that all of the parts are connected to the appropriate circuit points. A pictorial representation of these circuits is provided in Fig. 2C.

A Few Subtleties

You are probably wondering why the capacitors (sometimes wrongly called "condensers") have a curved line at one end and a straight one at the other end. The curved line indicates the end of the capacitor that goes to the terminal of lowest impedance or potential, such as circuit ground or the least positive of the two circuit points between which the part is installed. This applies mainly to polarized capacitors. Most of these parts are marked with a + symbol or may have a black band at the opposite end to indicate the negative terminal of the capacitor. This concept does not apply to disc-ceramic, mica and other nonpolarized capacitors, but the curve is always used in the symbol to show which end represents the low-impedance side of the circuit. Always pay close attention to the + symbols of capacitors: Hooking them up backward can cause them to short out or even explode!

Notice also that within the circular borders of Q1 and Q2 are arrows on the emitter line. When the arrowhead points toward the outer circle, the device is an NPN type, which requires a positive voltage on the collector terminal. If the arrows point inward toward the junction of the three lines, it signifies a PNP transistor, which needs a negative collector voltage. If you use the wrong transistor you may destroy it when voltage of the improper polarity is applied to it.

The arrowhead on R6 — the audio-gain control (sometimes called a "pot" for potentiometer) — tells us that the resistor value is variable by means of mechanical adjustment. In this case we would have the control mounted on the front panel of our equipment. Its shaft would be fitted with a knob to permit us to adjust the value of R6 when we wished to. If the adjustment were to be made only one time, then left in a preset position, we might install a trimmer pot at R6 (screwdriver adjust), and it could be installed right on the circuit board or chassis. Some hams call these controls "trimpots," but Trimpot® is a trade name, not a generic term.

J1 and J2 are jacks into which we may plug our outboard circuits or accessories, such as a microphone at J1. This electrical symbol is representative of a number of styles of jack. So just think of it as a connector of your choice — one that has a "hot" (center) terminal and a ground point (outer ring). It could be a phono jack, or one that a standard audio plug mates with. It could even be a coaxial-cable jack, if you wanted to use something that unusual for

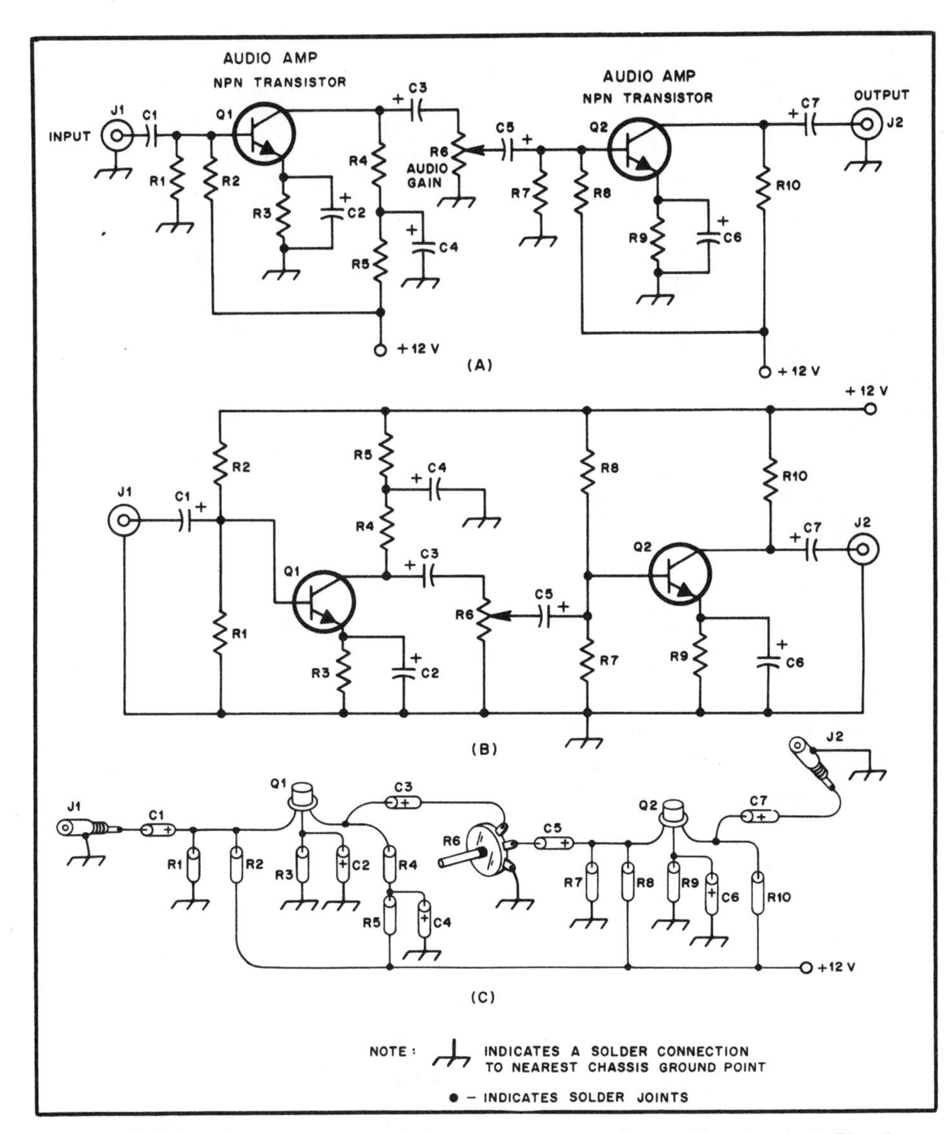

Fig. 2 — Examples of an identical circuit (A and B), drawn in different ways (see text). The pictorial representation at C is for the circuits shown in illustrations A and B. This shows how we can relate the drawing to the assembled circuit, which would normally be mounted on a circuit board or a metal chassis.

audio work! Examination of Fig. 1 will show that jacks with additional electrical contacts have a more complex symbol.

Notice that the symbol for ground in Fig. 2 looks like a rake. This is the proper symbol for *chassis ground* in a circuit. The earth-ground symbol of Fig. 1 is frequently misused by publishers for indicating chassis ground. Try not to be confused if you encounter disparity of this type: There is a significant difference between an earth ground and a chassis ground!

Voltage- and Ground-Bus Lines

I've noticed that one of the points of greatest confusion among beginners is how to configure the chassis-ground connections and the voltage-line network. In bygone days, when hams used wooden chassis, it was standard practice to run a ground-bus wire across the chassis. Each ground point in the circuit was then tied to this line by means of the shortest connecting lead possible. Other builders would return all ground connections for a single-circuit stage to a nearby common terminal, then route a lead from that point to the ground-bus wire. Although these techniques could still be applied, it is easier for us (and often better in terms of circuit performance) to bring each ground return to the metal chassis or circuit-board ground in the immediate area of the stage being wired. Not only does this impart a neater end product, it aids circuit performance (stability and reduced losses) when the ground leads are kept short and direct. The chassis or circuit board ground foil serves as the old-style ground bus when we do this. The "bottom line" here is to not worry about the maze of ground lines in the diagram. Simply make your ground connections short and direct near the related circuit elements.

Voltage-bus lines are treated like the old-time ground-bus conductors. That is, they are "floated" above ground on insulating terminals (or along specific voltage-bus foils on circuit boards). The various circuit points that connect to the voltage lines are connected by means of short jumper wires, or by the related components themselves (resistors, for example).

Circuit Direction

Another question that is asked frequently is, "Which way does a circuit run on a diagram?" The confused person means, does the first stage of a circuit start at the left or right of a drawing? Frankly, it makes no difference. Traditionally, for reasons I don't comprehend, a circuit has commenced at the left of the page and proceeded to the right. For example, considering a transmitter, the VFO or crystal oscillator would begin at the left of the sheet, followed by the intermediate stages, with the PA (power amplifier, or last stage) at the far right. Hams have developed the habit, as a consequence, of laying out the assembled unit from left to right also. I always did! But, it matters not how you lay out your project, provided you isolate one stage from another by reasonable physical separation, or by means of individual shield compartments. The last stage should never be placed alongside the input stages, lest unwanted feedback occur. The straight-line layout is the best method to adopt when in doubt.

Although it may not be apparent when examining a schematic diagram, we should always try to physically isolate the input and output components of a circuit stage from one another. Grouping them together will often cause feedback (output energy being fed back to the input circuit), which can cause a stage to self-oscillate, which renders the circuit useless. Some diagrams show a particular stage or stages enclosed in dashed lines. This indicates that that part of the circuit is contained in a shielded compartment to isolate it from the remainder of the circuit. A solid line around a circuit normally indicates that it is a separate module of a composite unit.

Potentiometers and Meters

We can't tell from the electrical symbol which end of a potentiometer (volume, tone, drive control, etc.) should be connected to ground. Many beginners have a problem with this: After wiring in the control, it operates backwards! For example, maximum volume occurs when the control is set fully *counterclockwise*. I understand this annoyance, for it used to happen to me!

Also, the circuit symbol for meters shows that one terminal is plus and the other is minus. But, which is which? Some meters have the polarity marked on the cases: Others bear no identification. Fig. 3 shows which end of a control should go to ground, and the meter drawing indicates which terminal is the positive one. The positive meter lug always connects to the circuit point of *highest* potential, as shown by the examples in Fig. 3. Incorrect

Glossary

base — the internal part of a bipolar transistor that controls the flow of current.
bus — a conductor of electrical current that carries a potential from one point in a circuit to another, such as positive or negative voltage, or ground.
capacitor — a device that stores dc energy but prevents its flow; permits the passage of ac energy, however.
cathode — negative electrode from which electrons flow in a stream inside a vacuum tube.
collector — in a bipolar transistor, the region through which the primary flow of charge carriers leaves the base. Generally, the output terminal of the transistor.
diode — a device having an anode and cathode, and which allows current to flow only one way.
disc-ceramic — a type of capacitor containing a ceramic dielectric (nonconducting material).
drain — a field-effect transistor electrode that supplies the amplified output signal in a grounded-source or grounded-gate hookup.
emitter — the element in a bipolar transistor that injects electrons into the base, which can be modulated by the base input signal.
encapsulated — a component that is embedded in a hard protective substance, or in a metal case.
feedback — ac energy that follows a path from one part of a circuit to another, intentionally or otherwise.
filament (heater) — in a vacuum tube, metallic wire heated by electric current; may serve also as the cathode in some tubes.
gate — part of an electronic device such as a field-effect transistor that controls the passage of current.
grid — in a vacuum tube an electrode that controls current flow.
hand key — a device used for sending Morse code.
impedance — the total resistance in an electrical circuit to the flow of alternating current at a specific frequency; expressed in ohms.
impedance, low — minimal resistance to ac.
integrated circuit — an electronics component that contains many individual transistors, diodes, capacitors and resistors and is sealed permanently in a single block or unit (unrepairable); usually referred to as an "IC" or "chip"; the various internal components are connected together or "integrated."
mica — an insulating (or dielectric) material found in nature; a mineral silicate.
oscillator — a circuit that generates a particular frequency.
plate — in a vacuum tube, the anode (positive element); in capacitors, the internal metal conductors.
polarized — a component that has positive and negative terminals marked on the case; the polarity is sometimes indicated by the shape of the part — each end being slightly different.
potentiometer — a variable resistor, such as a volume control.
resistor — a component that opposes the flow of current; available in a wide range of ohmic values and power ratings.
schematic — a diagram using electrical symbols that illustrates a circuit plan or "scheme."
semiconductor — an electrical component that is made from solid crystal materials, such as silicon or germanium; modern diodes, transistors and ICs are semiconductors; conductivity is intentionally poor compared to metal conductors.
source — the element in a field-effect transistor that supplies electrons; similar to the emitter in a bipolar transistor or the cathode in a vacuum tube.
transistor — a triode or tetrode semiconductor device that is capable of performing amplification, oscillation and control functions.
trimmer — an adjustable component, such as a capacitor or resistor; generally used for fine adjustment and left in a preset position.
vacuum tube — a device used to generate or amplify signals; can also be used as a rectifier or to perform control functions.
VFO (variable-frequency oscillator) — an oscillator whose frequency can be varied over a wide range by mechanical or electrical means; normally adjustable from the front panel of the equipment.

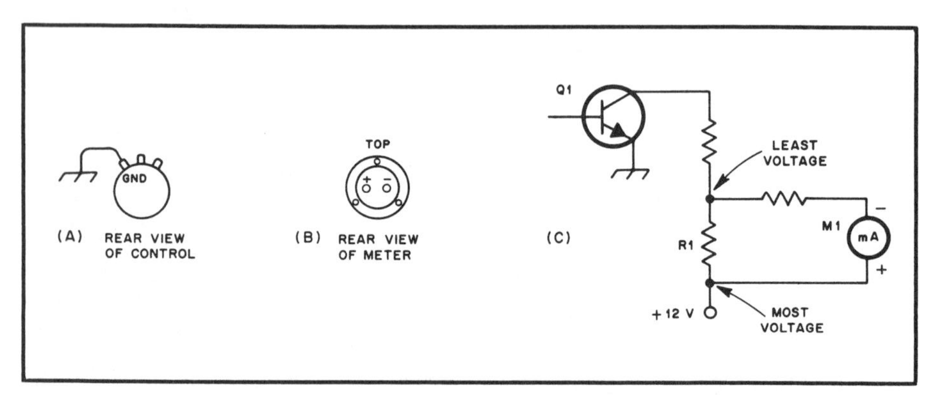

Fig. 3 — When wiring audio or tone controls, the ground end of the control is at the far left when viewing the control from the rear (A). Similarly, when viewing a meter from the back side (B), the positive terminal is at the left. The circuit at C shows that the negative terminal of a dc meter must be connected to the circuit point that has the lower of the two available potentials. The voltage drop across R1 in this example, caused by the current taken by Q1, makes the dc voltage lower at the top of R1 than it is at the low end of the resistor. This type of circuit can be used to monitor the current that Q1 draws.

polarization can destroy a meter at once. No one likes a meter with an S-shaped needle, jammed all the way to the left of the meter face! Ouch!

Some Final Words

The intent of this article is to help prepare you for the installments that follow in First Steps in Radio. How adept you become at following a schematic diagram easily and accurately will depend entirely on your tenacity in learning the symbology. Now is the proper time to apply yourself. This will make the lessons that follow a lot less difficult to digest. Practice drawing some simple circuits from memory. But, don't worry about the quality of your artwork. We aren't trying to follow in the footsteps of Rembrandt when drawing our diagrams; clarity is all that is required!

Understanding Resistors

Part 3: Without resistors we would be unable to build electrical circuits. What part do they play in a circuit? How are they rated? What do the color bands mean?

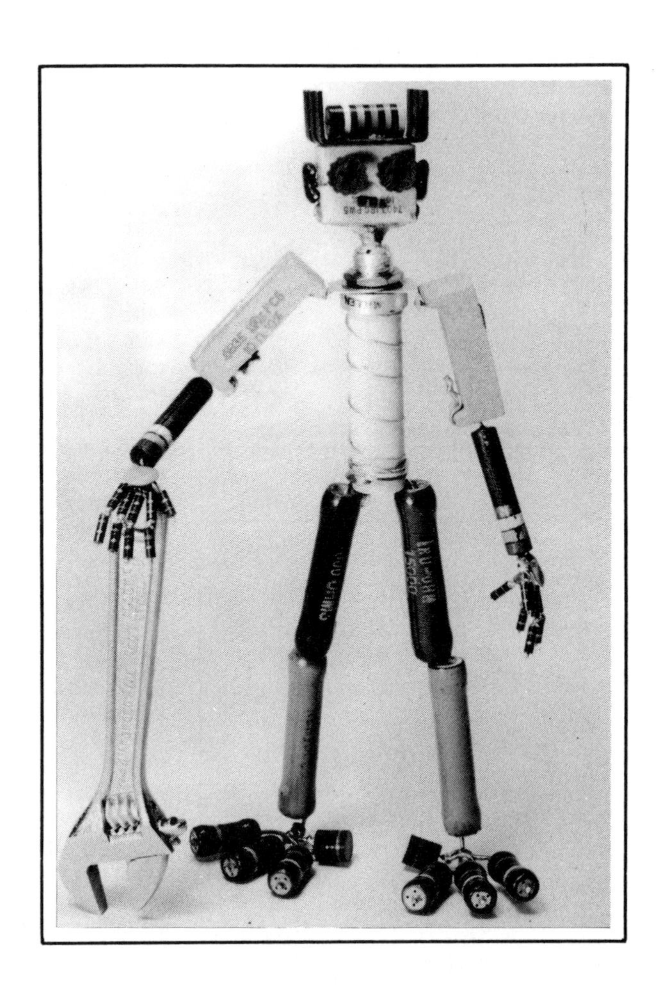

What is a resistor? Well, it is an electronic component that functions precisely as the name implies — it *resists* alternating or direct current. Resistors come in many sizes, shapes, power ratings and tolerances. Some have the value (resistance is specified in ohms) stamped on the case, while others have a group of color bands that help us to learn the resistance value. Let's learn more about resistors.

The Nature of Resistors

We can think of the resistor as an imperfect conductor. On the other hand, a perfect conductor would have *no resistance* at all. Therefore, if we had a test instrument that could accurately read ac or dc resistance to a finite value (zero ohms, in this case), the instrument would indicate zero ohms when the test set was arranged to read the resistance from one end of the conductor to the other (see Fig. 1). Perfect or nearly perfect conductors are necessary in many electronic circuits, but we also need to have poor conductors — namely resistors — in many parts of our radio circuits. This is where the resistor does its job.

In electronics work we usually measure the resistance of a material with an instrument known as an *ohmmeter*. Many hams and experimenters own a VOM (volt-ohmmeter) that is used for this purpose. A VOM also measures ac and dc voltage, and may include a function for measuring dc. Inexpensive VOMs can be obtained from Radio Shack, Heathkit® and similar outlets that sell components for experimenters. You should acquire a VOM for use in learning radio theory (lab experiments) and for design and repair work after you become experienced in Amateur Radio.

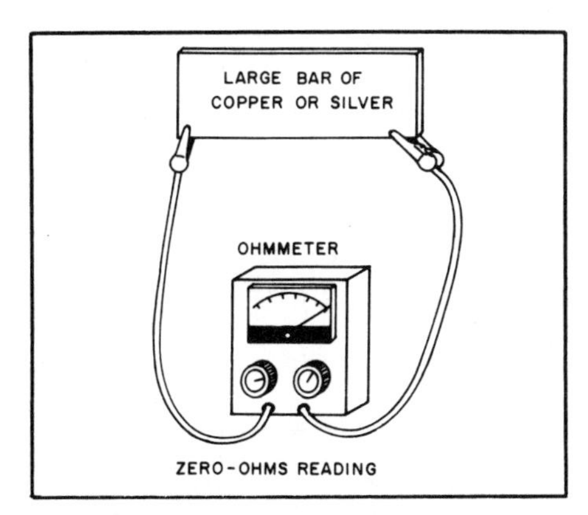

Fig. 1 — A perfect conductor would show a zero resistance. A large bar of highly conductive metal might represent a perfect conductor.

Power Classification

The greater the current that flows through a resistor the higher the power (wattage) rating must be. Resistors are available with ratings from as low as 1/8 watt to hundreds of watts. If the power rating of a resistor is too low for a particular circuit, it will get hot and burn out, sometimes quickly and other times gradually, depending on how much lower the rating is than the circuit application requires. When power is dissipated (as within a resistor) there will be heat. This is demonstrated clearly by an electric heater or toaster. The heating element in such appliances is a gigantic power resistor made from nichrome wire. This wire has a resistance that causes power to be dissipated as current passes through the wire. The wire glows from a red color to an almost yellowish color in some instances.

We could not tolerate having our radio-circuit resistors get that hot, so they are designed to operate cool or slightly warm to the touch. Choosing the correct power rating is, therefore, essential (more on this

Table 1

Resistor Color Codes†

Band Color	*Number (First Two Bands)*	*Zeros (Last Band)*
Black	0	—
Brown	1	0
Red	2	00
Orange	3	000
Yellow	4	0,000
Green	5	00,000
Blue	6	000,000
Purple	7	0,000,000
Gray	8	00,000,000
White	9	000,000,000

†Used on small carbon-composition units

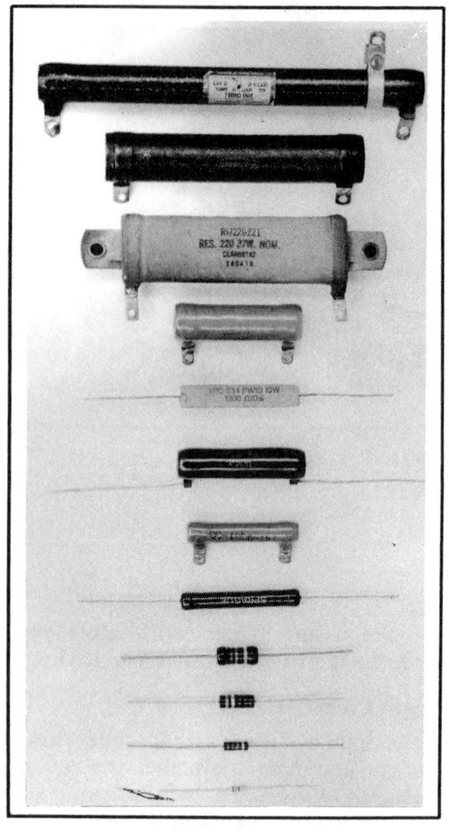

Fig. 2 — Photograph of various types of common resistors. The high-wattage types start at the top and the low-power resistors are at the bottom.

later). Fig. 2 shows a number of resistors of various wattage ratings. Low-power radio circuits (such as pocket-size transistor radios) use very tiny resistors (¼-watt sizes) because very little current flows in those circuits. On the other hand, we may find huge power resistors in large items of equipment, such as power supplies, that deliver large currents.

How to Read a Resistor Color Code

If we are to work with resistors we must learn how to determine their values from the color bands that are printed on them. Table 1 lists the colors found on resistors and shows what each color band represents numerically. You will want to memorize these numerical designators to be able to recognize and select them easily later on. There is usually a fourth color band on small carbon-composition resistors. It indicates the tolerance of the resistor in ohms — the percentage the actual resistance can vary, plus or minus.

Fig. 3 shows some examples of resistors with color bands, and provides the ohmic value of each. Remember that the term "k" means *thousand* and "M" stands for *million*. Thus, a 2.2-kΩ (kilohm — the omega symbol stands for ohms) resistor has 2200 ohms of resistance. Similarly, if the resistor is a 2.2-M (megohm) unit, the resistance is 2,200,000 ohms. Resistors are available with ratings from a fraction of an ohm to millions of ohms, but they come in *standard values* only. That is, resistors are not available for every possible ohmic round-number value.

Table 2 lists the standard values of primary interest to amateurs. If we need a special resistance value that falls between the standard values we can purchase, we must use a combination of resistors in parallel or series to obtain the needed value. More on that later. Alternatively, we may use a variable resistor (one for which the value can be changed by mechanical adjustment over a specified ohmic range). A volume control on a radio is an example of a variable resistor (also known as a potentiometer or "pot").

Physical Forms

Various formats are used in the manufacture of resistors. Some have wire leads (pigtails) that come out of the ends of the resistor bodies. Others have a tab at each end to which we may solder our circuit connections. The variety with tabs are called "power resistors" and are quite large. Some integrated circuits (ICs) contain microscopic arrays of resistors. Connection to those resistors is by means of the pins on the IC body.

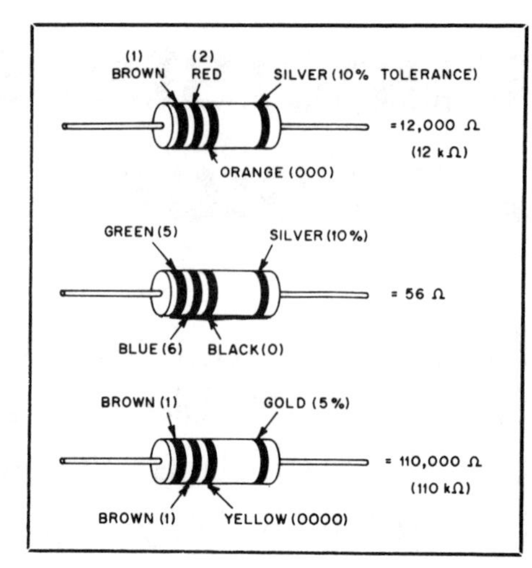

Fig. 3 — Three color-code examples to illustrate how to determine the value of a resistor that is banded.

There are many kinds of variable resistors. Some have sliders that make contact with the wire from which the resistor is made. As the slider is moved from one end of the resistor to the other, the effective resistance is changed. Panel-mounted variable resistors are used as volume and tone controls, as well as for a host of other functions, such as adjustment controls on TV sets. Other circuits contain variable resistors that must be adjusted by means of a screwdriver. These are called "trimmer resistors" or "Trimpots," which are generally set for a specific resistance just

Table 2

Standard Resistance Values

Resistors with ± 10% tolerance are available only in values shown in bold type. Resistors with ± 5% tolerance are available in all values shown.

Ohms

1.0	3.6	**12**	43	**150**	510	**1800**	6200	**22000**	75000
1.1	**3.9**	13	**47**	160	**560**	2000	**6800**	24000	**82000**
1.2	4.3	**15**	51	**180**	620	**2200**	7500	**27000**	91000
1.3	**4.7**	16	**56**	200	**680**	2400	**8200**	30000	**100000**
1.5	5.1	**18**	62	**220**	750	**2700**	9100	**33000**	110000
1.6	**5.6**	20	**68**	240	**820**	3000	**10000**	36000	**120000**
1.8	6.2	**22**	75	**270**	910	**3300**	11000	**39000**	130000
2.0	**6.8**	24	**82**	300	**1000**	3600	**12000**	43000	**150000**
2.2	7.5	**27**	91	**330**	1100	**3900**	13000	**47000**	160000
2.4	**8.2**	30	**100**	360	**1200**	4300	**15000**	51000	**180000**
2.7	9.1	**33**	110	**390**	1300	**4700**	16000	**56000**	200000
3.0	**10.0**	36	**120**	430	**1500**	5100	**18000**	62000	**220000**
3.3	11.0	**39**	130	**470**	1600	**5600**	20000	**68000**	

Megohms

0.24	0.62	1.6	4.3	11.0
0.27	**0.68**	**1.8**	**4.7**	**12.0**
0.30	0.75	2.0	5.1	13.0
0.33	**0.82**	**2.2**	**5.6**	**15.0**
0.36	0.91	2.4	6.2	16.0
0.39	**1.0**	**2.7**	**6.8**	**18.0**
0.43	1.1	3.0	7.5	20.0
0.47	**1.2**	**3.3**	**8.2**	**22.0**
0.51	1.3	3.6	9.1	
0.56	**1.5**	**3.9**	**10.0**	

once, then left in that position. Fig. 4 shows a number of variable resistors.

Putting the Resistor to Work

Let's imagine that we built a small transistorized audio amplifier designed to increase the output from a microphone. We would need some resistors to perform electrical tasks within the circuit. The diagram in Fig. 5 illustrates our use of resistors. The illustration at drawing A is a refresher of sorts on how to read a diagram (see Part 2). It shows the physical aspects of our little microphone amplifier. Examine the schematic diagram at B of Fig. 5. Note that at the top end of R4 we have a lower voltage than is found at the battery terminals. That is because R4 is a resistor, and when the transistor (Q1) draws current through R4 it will cause what is known as a *voltage drop*. The higher the current flow, or the greater the resistance of R4, the greater the voltage drop across the resistor.

This can be used to advantage in many circuits where the battery or power-supply voltage is too high for a particular transistor, tube or IC. The proper value of resistor is used to ensure that the transistor is protected from excessive voltage or current. Too much voltage (and the increased current) can cause the transistor to overheat and be destroyed, or the excessive voltage might puncture the inner elements of the transistor and destroy it.

In order for us to select a correct value of resistance for R4, we need to know the amount of current in that branch of our circuit. That exercise is beyond the purpose of this discussion, but we mention it now for tutorial purposes. Once we know the current value in such a circuit (we'll assume it is 1 mA in Fig. 5), we can choose a resistor value to provide the desired operating voltage. We will use Ohm's Law, which shows the relationship between resistance, voltage and current in simple algebra:

$$R = \frac{E}{I} \text{ ohms} \qquad \text{(Eq. 1)}$$

where E is the desired voltage drop, and I is the circuit current in amperes (note that 1 mA equals 0.001 A). Hence, if we have a 9-V battery and desire 4.7 volts at the collector of Q1 (Fig. 5B), our resistor must drop 4.3 volts. Its resistance will be determined by

$$R = \frac{4.3 \text{ V}}{0.001 \text{ A}} = 4300 \text{ ohms} \qquad \text{(Eq. 2)}$$

If we subtract 4.3 (the voltage drop across R4) from 9 volts, we have 4.7 volts at the collector.

Fig. 4 — Photograph of assorted variable resistors. Some can be adjusted by means of a knob, while others require a screwdriver to change the effective resistance value.

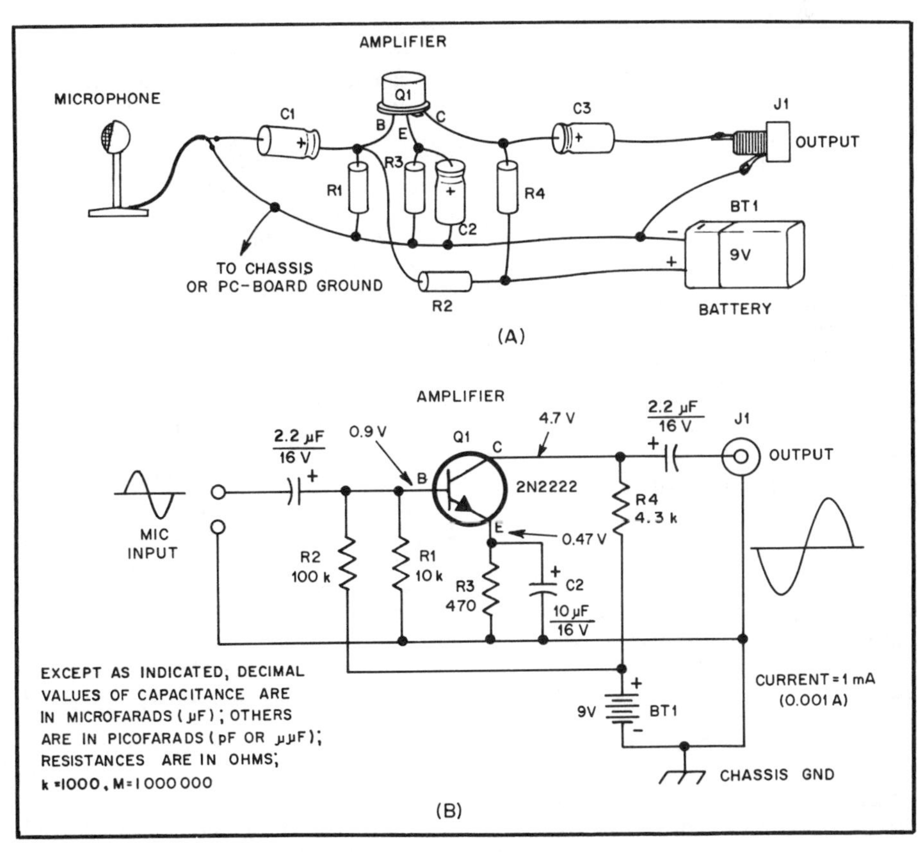

Fig. 5 — Pictorial (A) and schematic (B) examples of a simple one-stage audio amplifier. This set of examples is presented for text-discussion purposes.

R4 serves still another purpose in our circuit. Since it resists the passage or flow of dc and ac, it will hold back our amplified voice signal (composed of ac energy) and prevent it from being lost into ground via the battery. Instead, the audio energy is directed to the output jack (J1) through capacitor C3. If R4 were too low in resistance we would lose a large part of the audio signal before it reached J1.

R1 and R2 are used at the input of our Fig. 5 circuit for the purpose of establishing a small operating voltage (approximately 0.9 V) at the base of Q1. Those resistors also isolate the signal from our microphone so it is routed to Q1 rather than to ground via BT1.

Our circuit needs a small voltage at the emitter of Q1, so we are using R3 to develop what is called *self bias* (emitter bias). The 0.001-A current of the transistor also flows to ground through R3. This creates a

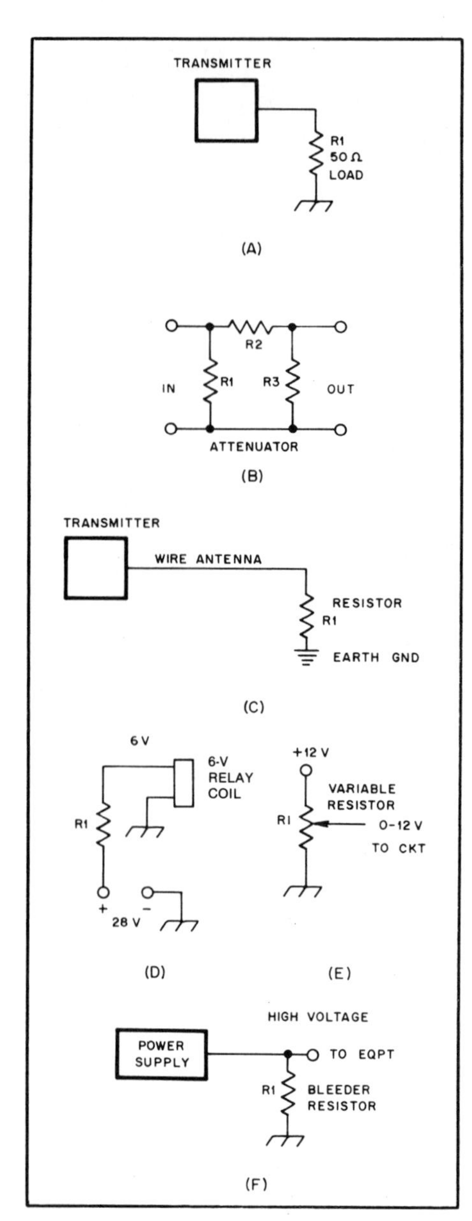

Fig. 6 — Various applications for resistors (see text).

voltage drop across the resistor, which in this case is 0.47 V. This can be calculated from another form of Ohm's Law:

$$E = IR$$
$$= 0.001\text{ A} \times 470\ \Omega = 0.47\text{ V} \quad \text{(Eq. 3)}$$

where I is in amperes (A) and R is in ohms.

If you have a voltmeter available I suggest you obtain the parts for the circuit in Fig. 5 and tack it together for experimental use. Try various resistance values at R4 to see how the collector voltage at Q1 changes. Of course, as the resistance is made greater, the current drawn by Q1 will fall. But, once you measure the voltage drop across R4 you will be able to calculate the current of Q1. Since Ohm's Law is the basis of electrical work, and may appear on license exams, you should practice your calculations now. You can learn what the current flow is from still another version of Ohm's Law:

$$I = \frac{E}{R}\text{ amperes} \quad \text{(Eq. 4)}$$

where E is the drop across the resistor in volts and R is the resistor value in ohms. Furthermore, if you build the test circuit neatly on a printed-circuit or "perf" board, you can try it with your microphone by placing it between your mike and the input of an audio hi-fi amplifier.

I believe strongly in "learning by doing." I hope you will get involved with the simple lab experiments suggested in this series. They will bolster your "book larnin."

Some Other Uses for Resistors

The applications for resistors are so numerous, and at times detailed, that we could fill an entire book discussing them all. But for the sake of brevity let's examine a few examples of where we might apply resistors in routine amateur work.

Fig. 6A shows a transmitter to which we have connected a 50-ohm resistor (R1). This is called a *dummy load* or *dummy antenna*. If R1 is the same resistance as the transmitter output (usually 50 ohms), and if it can safely handle the transmitter output power, we may use R1 in place of an antenna during transmitter tests. We thereby prevent our signal from going out over the air and possibly interfering with another amateur.

Fig. 6B shows three resistors used in an *attenuator* circuit. Attenuators can be designed to reduce the power by almost any amount we desire. The example shows a circuit that will reduce the input power to a desired output power level. The amount of power reduction (attenuation) depends on the resistance values selected for resistors R1, R2 and R3. In other words, if we had a low-power transmitter we wanted to use to drive a high-power amplifier, but the small transmitter put out too much power, we could use an attenuator. It would be placed between the transmitter and amplifier. If, for example, the attenuator in Fig. 6B cuts power in half, we would get 30 watts into the amplifier if the transmitter put out 60 watts.

Resistors are sometimes used in antenna systems, as shown in Fig. 6C. R1 can be used to make the antenna present a particular resistance to the transmitter and receiver. An antenna of this type is called a *terminated antenna,* because the resistor is used at the far end (termination).

Sometimes we hams buy surplus relays for our projects. They may have the wrong voltage rating for the power supply we have on hand. Fig. 6D shows how we might lower the relay operating voltage if it requires a lower potential than that of our power supply. To find the value of R1 we can measure the resistance of the relay coil with an ohmmeter, then apply Ohm's Law in accordance with the voltage drop needed.

Earlier in this article we talked about variable resistors. An example of one is given in Fig. 6E. The resistor has a movable contact that can be varied for any voltage from 0 to 12.

Finally, in Fig. 6F we see a resistor being used as a *bleeder.* Power supplies that provide dangerous voltage potentials (hundreds or thousands of volts) are equipped with bleeder resistors. Your ham license exam may have a question about this. The resistor permits the power-supply voltage to trickle or bleed off slowly (seconds) when the supply is turned off. This protects the operator against an accidental shock (which could be lethal) from the charge stored in the filter capacitors. R1 has a sufficiently high resistance to prevent it from taxing the power supply (drawing excessive current) during normal operation.

The Wattage Rating of Resistors

Each resistor we use must be chosen in accordance with the power that will be dissipated within it. If it isn't, we can burn up a resistor rather quickly! Resistors come with various wattage ratings, and for most low-current, low-voltage operations (such as in transistor radios) we will use ¼- or ½-watt units. The wattage rating of a resistor signifies the *maximum* safe power it will dissipate without changing value or burning out. As a safety margin it is wise to use the next higher rating than the circuit demands. In other words, if ½ watt of power was dissipated in a resistor, a ½-watt unit would be hot to the touch and you'd want to use a 1-watt unit.

We can learn the power consumption in a circuit branch if we know any two of the voltage, current or resistance values.

$$P = E \times I$$
$$= I^2R$$
$$= \frac{E^2}{R} \quad \text{(Eq. 5)}$$

where P = power in watts, E is in volts, I is current in amperes and R is in ohms. Thus, if the branch of our circuit that contains a resistor has a current flow of 50 mA (0.05 A) and the resistor is 470 ohms, the power dissipated in the resistor will be 1.175 watts, from

$$P = I^2R = 0.05^2 \times 470\ \Omega = 1.175\text{ W} \quad \text{(Eq. 6)}$$

This tells us that a 2-watt resistor should be installed in that part of the circuit.

Resistor Combinations

At the start of our discussion I mentioned combining resistors in series or parallel to obtain special values of resistance. How might we do this and know the resultant value of resistance? Simply by doing a bit of basic math with a calculator or slide rule, or by longhand.

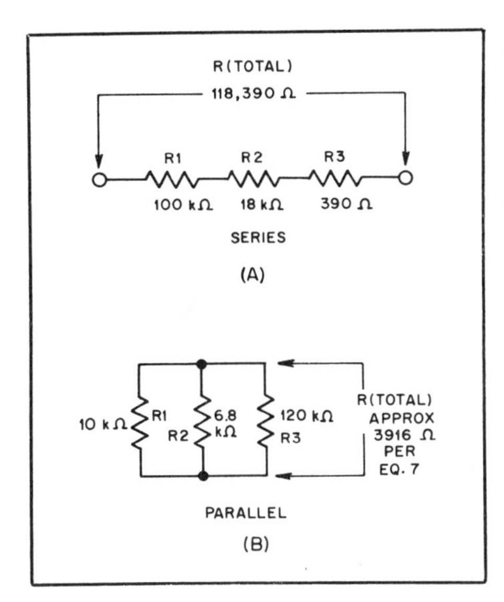

Fig. 7 — Resistors in series result in the combined ohmic value of the string (A). When resistors are wired in parallel (B) we must use Eq. 7 to learn the resultant total resistance.

Glossary

bias — a voltage, normally on the input lead of an active device (tube or transistor), to make the active device operate in a desired region. Bias sets the resting voltage when no signal is present.

parallel — a way of connecting two or more components together in a side-by-side manner. The current is divided between the two or more branches formed. For example, the following two resistors are connected in parallel:

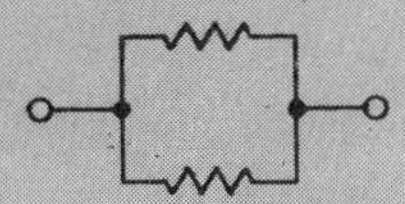

relay — a switch that can be remotely controlled by an electrical signal. The control signal goes to a coil that magnetically pulls an armature, which in turn causes the switch to move from its normal position.

series — a way of connecting two or more components together in a string so the same current goes through each. For example, these two resistors are connected in series:

When we connect resistors in series (Fig. 7A) we merely add the values of the individual resistors. But, when we place the resistors in parallel we must use Eq. 7.

$$R(\text{total}) = \frac{1}{\frac{1}{R1} + \frac{1}{R2} + \frac{1}{R3} \cdots} \text{ ohms (Eq. 7)}$$

Fig. 7B shows a combination of series resistors and the net value of resistance, as determined from Eq. 7. When resistors are used in this manner they occupy considerably more room in the circuit than if a single unit were employed. But, using series or parallel combinations is often necessary to obtain a critical value of fixed resistance.

Special resistance values can often be obtained from *precision resistors* that can be purchased on special order. They are costly and are thus not apt to be a product you will ever want to buy. Also, close-tolerance resistors (1%) are available at increased cost from large parts distributors. For most practical applications, precise values are unnecessary — both to the circuit and to your pocketbook.

Final Comments

I hope you now have a basic understanding of what resistors are and why they are necessary. Let me encourage you to obtain some small hand tools, a soldering iron, a VOM and some rosin-core solder. This will enable you to do experiments as we progress through this book.

Getting to Know Capacitors

Part 4: Along with resistors, which we explored last time, capacitors are an integral part of radio electronics. Don't bypass this installment.

What purpose do capacitors serve? We could compile a long list of functions if time and page space permitted. But, for this discussion we will examine their most common applications. You may have heard people refer to them as "condensers." That misnomer has been popular since the beginning of radio, but it does not describe the function of a capacitor. A retort in a chemistry lab is a *condenser,* since it converts vapor to liquid, but a capacitor can't function that way. Rather, it stores dc energy, but permits the passage of ac energy. In other words, the device has a *capacity* for storing energy, the magnitude of which is expressed in farads (F), microfarads (μF) or picofarads (pF). Capacitors are rated in some parts of the world in nanofarads (nF) as well. A capacitor is a component that consists of two electrodes separated by a dielectric (an insulator), such as air, or some solid material.

The *dielectric* material of a capacitor is the insulating medium between the electrodes, or *plates.* Generally, it is air for variable capacitors (sometimes called tuning, trimmer or padder capacitors). The insulation in fixed-value capacitors may consist of treated paper, mica, glass, polyethylene, polystyrene, Mylar or a host of other substances. Some variable capacitors (movable plates) have a solid dielectric between the plates. Smaller trimmer or padder capacitors have thin sheets of mica between their plates.

Each type of insulation has a different characteristic (factor) with regard to the voltage it can accommodate for a given thickness before breakdown (puncture). The dielectric factor (ϵ) also determines the amount of capacitance for a specified plate spacing and area. Dry air and helium are considered to be the best dielectrics for capacitors that must be used in high-voltage circuits.

Some Names and General Descriptions

Simply, we can think of a capacitor as a device that permits the flow of ac energy, but blocks the passage of dc energy. A basic capacitor is illustrated in Fig. 1A. The more plates we add, the greater the effective capacitance for a given plate spacing (Fig. 1B). The greater the plate spacing, on the other hand, the lower the capacitance and the higher the safe voltage rating. In the case of air-dielectric and other capacitors, an arc occurs between the plates when the voltage is too high for the insulation.

We will hear about and frequently use what is known as a "paper," or "tubular" capacitor. This type consists of many layers of treated paper or other insulating

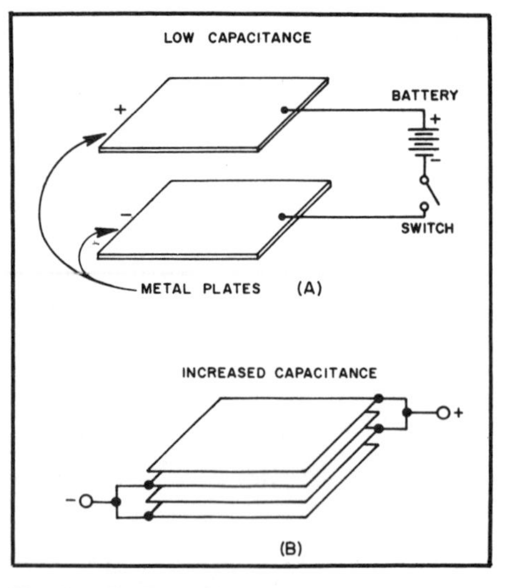

Fig. 1 — Example of a simple capacitor (A) that has two electrodes, or plates. A multiplate capacitor is shown at B. It has greater capacitance than the example at A.

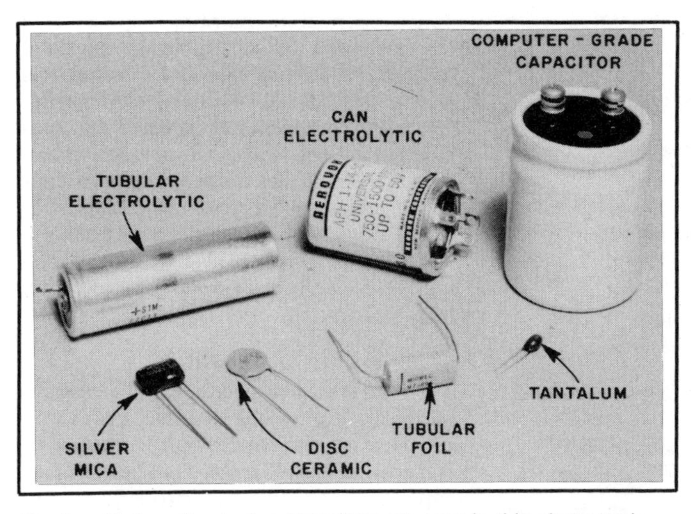

Fig. 2 — Various fixed-value capacitors are seen in this photograph.

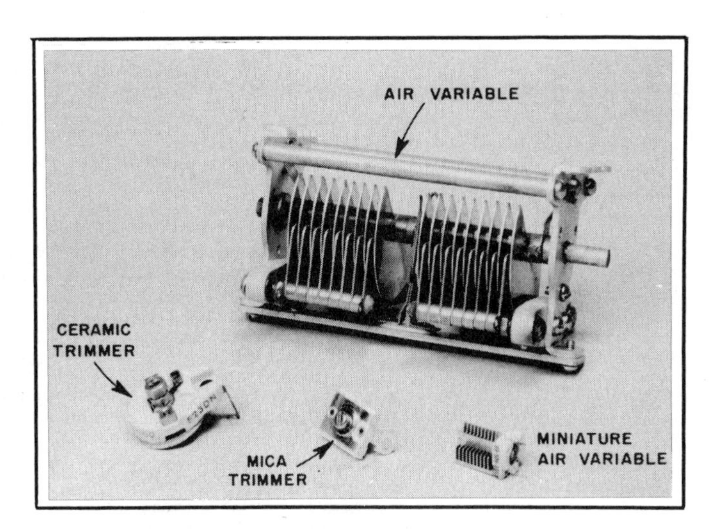

Fig. 3 — Examples of variable capacitors.

material, rolled into a cylinder. The surface of the insulation has a metallic electrode affixed to it. Two such electrodes are used in parallel, and each is insulated from the other. A wire lead is connected to one electrode, and another such lead is attached to the remaining electrode. If the capacitor is a polarized type, one terminal is marked with a minus sign (−) and the other has a plus sign (+). Some of these capacitors, when polarized, lack the plus and minus signs. Instead, there is a black band around one end of the capacitor to indicate the negative terminal. By rolling the electrodes and insulation into a tight cylinder it is possible to obtain large amounts of capacitance in a small package. This is not practical for variable capacitors.

You will hear also about "oil-filled" capacitors. These are designed for high capacitance and high voltage. The plates are insulated by an oil that has a high dielectric factor. We will find these units mainly in high-voltage power supplies.

The highest capacitance ratings can be found in what many people call "computer-grade" capacitors. These are large cylindrical units that contain a solid dielectric. They are encased in an aluminum cylinder. It is not unusual to find them with ratings of 50,000 μF or greater, and with dc-voltage ratings as high as 450.

Another type of high-capacitance device is the "tantalum" capacitor. These miniature low-voltage devices are ideal for use in printed-circuit-board assemblies because of their small size. They are considered to be high quality units, and have ratings well into the high-microfarad region.

Still another capacitor is the "vacuum" type. These units have the air evacuated from them at the time of assembly. They are built into glass containers with metal end caps. Because of this lack of true dielectric material (a vacuum), they can accommodate tremendous voltage levels without breaking down. They are expensive and large, but many amateurs use them in special circuits. Vacuum variable capacitors are also available as replacements for air variable capacitors.

The most common of the fixed-value capacitors is the "disc ceramic." They are disc-shaped and have two wire leads coming from them. Most of these capacitors look like pieces of gum that someone has stepped on. Ceramic is used for the insulating medium. They are small and especially convenient for use on etched-circuit boards. Values as great as 0.1 μF are common, and voltage ratings as high as 1000 or greater are available for the lower-capacitance units.

Another common capacitor is the "mica" type, named for its dielectric material. The plates of one variety are coated with silver, hence the name "silver mica." These are considered to be high quality components with reasonably good tolerance in terms of the marked values. They are often preferred to disc-ceramic capacitors when values less than, say, 2000 pF are needed in circuits that must be stable (such as in oscillators and filters for radio frequencies). They are not affected by heat as much as some other kinds of low-capacitance units. The effects of heat can be seen in a gradual change in capacitance value. Cold temperatures have a similar effect, but in an opposite direction with respect to net capacitance at a given instant. Some ceramic capacitors are specially manufactured to serve as "compensating" capacitors for various conditions of heat or cold. They are rated with particular "temperature coefficients."

You will use many capacitors that are called "electrolytics." These are designed for high values of capacitance (as are computer-grade and tantalum capacitors), and they have polarity marks on them. Electrolytic capacitors are used in power supplies (for ripple filtering) and in audio circuits as coupling and bypassing units.

Fig. 2 shows a collection of fixed-value capacitors. Variable capacitors appear in Fig. 3. To summarize this section of our lesson, we should remember that the farad (F) is the basic unit of capacitance. The μF is 10^{-6}F; the nanofarad is 10^{-9}F; the picofarad is 10^{-12}F. Learning these relationships will help us to perform pertinent mathematical exercises when we become proficient enough to start designing our own circuits.

Selecting and Using Capacitors

Some of you may have concluded after looking at schematic diagrams in *QST* that following or building a circuit is next to impossible. Not so! You need not fear the use of capacitors or resistors. Many beginners to electronics think they must use the exact item called for in the parts list; this is seldom a requirement. It is important, however, to know which types of capacitors are best for a certain kind of circuit. Table 1 provides a useful generalization concerning capacitors and the kinds of applications they are best suited for.

The value specified is not critical in many circuits. The notable exception is seen when a capacitor is part of a frequency-determining circuit (such as tuned circuits

Table 1
Common Capacitor Types and Applications

Application	*Capacitor Type*	*Frequency Range*
Radio-frequency circuits	Disc ceramic, silver mica, tubular ceramic, polystyrene and air or ceramic variables. Mica trimmers and teflon trimmers are also suitable.	50 kHz through VHF (see text).
UHF and microwave circuits	Ceramic chip capacitors. Glass piston trimmers and some small air-dielectric trimmers. Air disc variables are also used.	150 through 1296 MHz, generally.
Audio and very-low-frequency (VLF) circuits	Tantalum, electrolytic, tubular, Mylar, oil-filled and all of the above.	10 kHz through 500 kHz, generally.
Power supplies and voltage regulators	Tantalum, oil filled, computer grade, electrolytic and tubular paper.	25 Hz through 3 kHz.
Tuned circuits at radio frequencies	Air variables, mica trimmers, ceramic trimmers, glass piston trimmers and tuning diodes.	50 kHz through VHF and higher.

in transmitters and receivers), or when they are used in what is known as a "timing circuit." The voltage rating is *always* important, though. If a specific voltage rating is listed, it's alright to use a unit with a higher voltage rating; but don't install a capacitor that has a lower voltage rating.

In circuits where capacitors are used for bypassing or coupling between stages, we can usually get by with values other than those specified. For example, if a transistor emitter is bypassed to ground with a 10-μF, 6-V capacitor, we can safely use, say, a 15-μF, 25-V unit, if that's all we have on hand. Similarly, if a transmitter circuit calls for a 100-pF ceramic coupling capacitor, we may substitute a 150-pF silver-mica unit without a significant change in performance. Or, if we happen to have only a 68-pF part on hand, we might use it as a substitute.

Depending on how they are constructed, some capacitors have an unwanted characteristic known as "stray inductance." What we are seeking in a capacitor is *pure capacitance*. We do not want the elements of resistance and inductance to be present, for they affect the quality of the capacitor — especially at the higher frequencies. Unfortunately, all capacitors have resistance and inductance associated with them. Some styles are much worse than others; therefore, we should always select the units we use in accordance with Table 1. This will help to minimize the effects of the resistance and inductance that are present.

A component I did not mention earlier is called a "chip capacitor." These specially manufactured small parts have no wire leads to cause unwanted inductance. They are square or rectangular in shape and have silver-plated electrodes. The dielectric is ceramic. You can actually solder a chip capacitor directly to the copper elements of a circuit board. This keeps the connecting leads of the circuit very short. Chip capacitors are used mainly at very-high frequencies (VHF) and above. Conventional fixed-value capacitors will not function as true capacitors in some VHF circuits. The main problem associated with chip capacitors is high cost and limited availability for ham use. As an example of what we are discussing here, a silver-mica capacitor may have a marked value of 56 pF. At 3.5 MHz it will function in accordance with that marking, since stray resistance and inductance will be too small to be significant at that low a frequency. But, at 146 MHz (VHF), it might appear as a 220-pF capacitor because of the combined effects of stray resistance and inductance. To minimize these effects, the leads of all capacitors used in a radio-frequency circuit should be cut as short as possible, still allowing ample lead length for soldering. In audio and power-supply circuits, this is not a requirement.

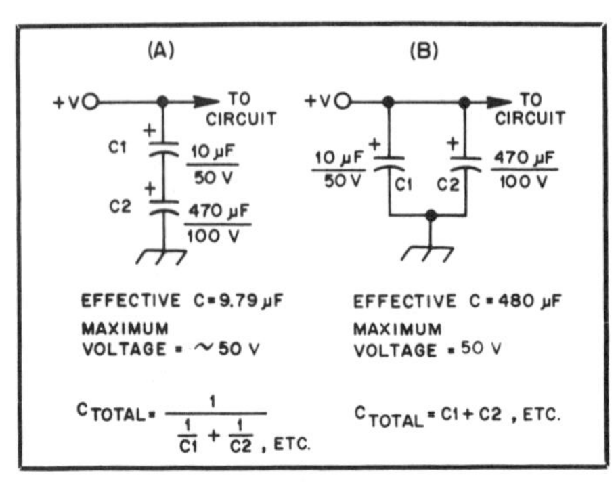

Fig. 4 — Capacitors in series are shown at A. The example at B shows capacitors in a parallel arrangement (see text).

Combinations of Capacitors

When we use capacitors in series or in parallel, we get the opposite effect than is experienced with similar configurations of resistors (Fig 4). That is, the net value of capacitance for two capacitors hooked in parallel is the capacitance of one plus that of the other. Conversely, when we use them in a series hookup, we end up with slightly less capacitance than the marked value of the smallest one. The voltage rating of parallel-connected capacitors is that of the unit having the *lowest* voltage rating. Series-connected capacitors divide the applied voltage according to the magnitude of their capacitance. The voltage across each capacitor is proportional to the total capacitance divided by that of the individual capacitor. It is best to use capacitors of the same voltage rating to avoid problems. We will often find equalizing resistors connected across the various capacitors in a series combination. This helps to ensure an equal division of operating voltage across each unit.

Capacitors in an Actual Circuit

Fig. 5 shows a typical two-transistor circuit for use as an audio amplifier. Examples of bypass and coupling (blocking) capacitors are given. C1, C2 and C7 are coupling capacitors. Audio energy is able to pass through them, but dc voltage is blocked by the capacitors. This is just like a gate for the audio (ac) voltage, but it

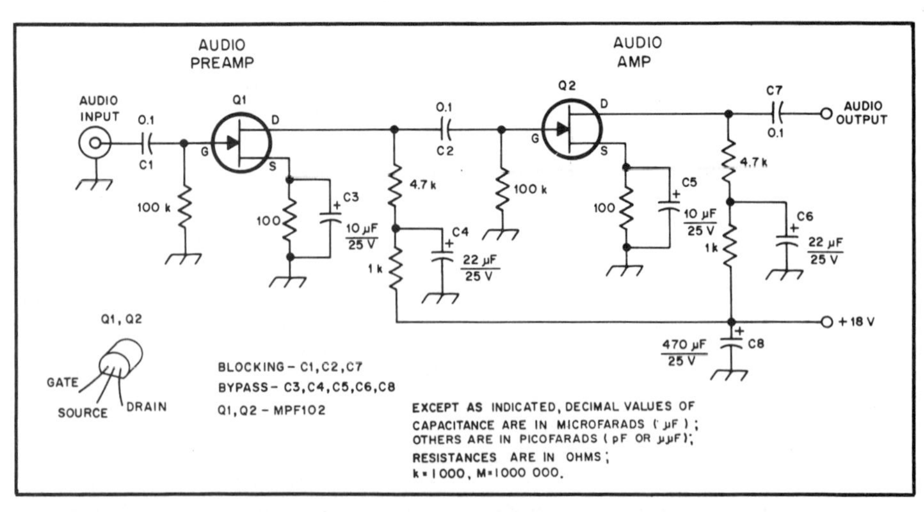

Fig. 5 — Illustration of a simple audio-amplifier circuit that uses coupling and bypass capacitors. Their functions are discussed in the text.

becomes a "brick wall" for the dc voltage. The ac voltage must pass through that branch of the circuit in order for us to have the benefits of audio amplification. But, if dc were permitted to pass through the same branch of our circuit, too much voltage would appear in the wrong places. The transistors could burn out, or amplification would not occur.

C3, C4, C5, C6 and C8, on the other hand, are bypass capacitors. We want the ac voltage to be directed to ground at these points. So, the capacitors permit this to happen while preventing dc voltage from going to ground. In other words, we are bypassing unwanted ac (audio) energy away from the circuit elements to which the capacitors are attached.

You will notice that C3, C4, C5, C6 and C8 have the polarity (+) indicated. Always be sure to hook the + sign to the plus-voltage line. The negative ends of those capacitors must go to ground. Otherwise, they could become hot, or even explode! I have had a few electrolytic and tantalum capacitors blow apart because I hooked them up backward. The sound is not unlike that of a gun going off! *Use caution.*

The polarized capacitors of Fig. 5 can be of the electrolytic, tantalum or computer-grade types with respect to satisfactory performance. The electrolytic units, preferred by most hams, will cost the least.

Still Another Type of Capacitor

A number of semiconductor diodes are available for use as "electronic capacitors." Such a device is shown in Fig. 6, as D1, which is the designator for a diode. In this circuit, the diode functions as a variable capacitor to change the frequency of the tuned circuit that includes L1 and C1. As the operating voltage (dc) is varied at the diode terminal by means of R1, the internal capacitance of the diode changes. R1 would therefore become our panel-mounted tuning control. We could use it with a dial that showed the frequency versus the setting of R1. This type of tuning element has a number of trade names connected with it, such as Varicap® and Epicap®. It is known also as a *varactor* diode. Many FM and TV sets use these diodes for tuning in the stations.

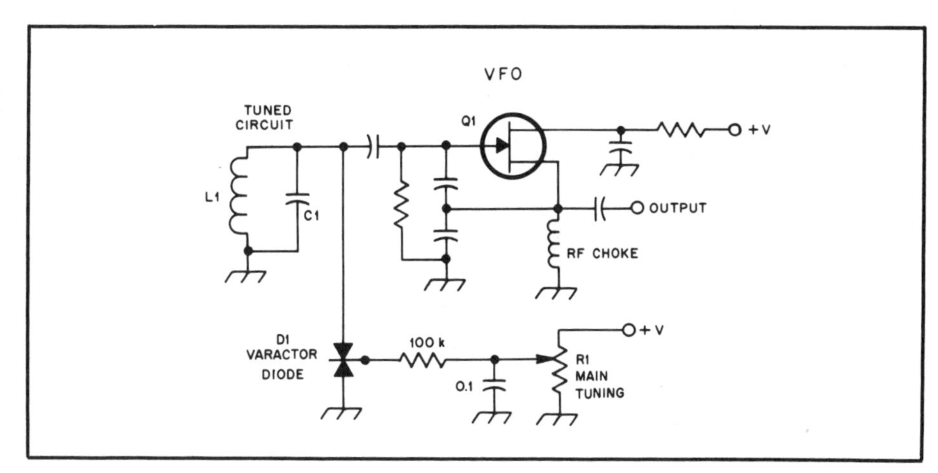

Fig. 6 — D1 in this circuit is an electronic capacitor. It is a semiconductor diode with two anodes and one cathode. You may learn how it works by reading the text.

Glossary

dielectric — an insulating medium. A nonconductor of direct current. Air, mica and ceramic are examples of dielectric materials.

electrode — a conductor that is used to establish an electrical connection to a nonmetallic part of a circuit. The plate cap on a vacuum tube is one example. A battery terminal is another.

electrolytic — an action caused by electrolysis. *Electrolyte* is the medium needed to make a device electrolytic. It is a nonmetallic conductor in which current is carried by the movement of ions. Such ionic conductors are found in wet batteries and electrolytic capacitors.

filtering — the electrical process of removing unwanted signals, ac frequencies or dc from incoming energy. Filters are used in receivers to permit the passage of desired signals while restricting the passage of unwanted, interfering signals. They are used also in transmitters to ensure purity of the output signal. Filters are used in power supplies to remove the ac ripple energy from the desired dc output voltage.

microfarad — a numerical expression for a portion of the farad, which is the basic unit of capacitance. Microfarad, expressed also as μF, is 0.000,001 farad, or 10^{-6} farad. For example, a 100-μF capacitor is 0.0001 farad.

padder — a variable capacitor, generally of small physical size and relatively low capacitance range, adjusted by means of a screwdriver or other small tool. Normally used to tune a particular circuit to a specific frequency. Sometimes padder capacitors are referred to as *trimmer* capacitors.

picofarad — abbreviated pF, a numerical part of the farad. It is 0.000,000,000,001 farad, or 10^{-12} farad.

stray inductance — unwanted inductance in a circuit. A piece of hookup wire or a printed-circuit board foil will introduce a certain amount of stray inductance. Too much stray inductance can spoil circuit performance. Becomes more pronounced as the operating frequency is raised.

tantalum — a substance used in the manufacture of certain types of capacitors. The size of the tantalum bead determines the capacitance of an individual unit. Tantalum capacitors can be made much smaller than electrolytic units of the same value.

varactor — variable-reactance diode. A semiconductor diode that functions as a voltage-variable capacitor. The capacitance of a varactor depends on applied voltage amplitude and polarity. These devices are used in place of mechanical tuning capacitors in many circuits.

very high frequency — known also as VHF. The band of frequencies that lies between 30 and 300 MHz. The 6- and 2-meter bands are examples of amateur VHF bands.

What Have We Learned?

Although we have only covered the basics of capacitors, we have discovered that they are a "must" in most electronics circuits. They can be used for many purposes, and they come in many shapes, sizes, ratings and types. Their greatest use is for bypassing (filtering) and coupling in radio circuits.

We can summarize additionally by recalling that each capacitor has a particular maximum voltage rating that we must pay close attention to. We need also to observe the polarity marked on some types. Most small capacitors are not polarized, which means we can hook either end to a circuit point.

It is not mandatory that we use the exact capacitor called for in a construction project. For the most part, we can use a value that is reasonably close to the indicated one, and no problems should be experienced. We can also substitute some kinds of capacitors for others when the need arises, consistent with the data in Table 1.

We may obtain specific values of capacitance by using two or more capacitors in series or parallel. This is handy when we do not have a particular value available for a project. A happy amateur is usually one who is an inveterate experimenter. Parts substitution is part of the game. If your circuit calls for a 100-pF variable capacitor, don't be afraid to use a 150-pF unit. It will simply extend the effective tuning range of the circuit. Using more, rather than less, is a good rule of thumb, for too little capacitance in a variable unit will restrict the tuning range. Many hams pull the plates (vanes) from a too-large variable capacitor until the maximum-capacitance value is close to the prescribed one. I've done it many times.

An Introduction to Coils and Transformers

Part 5: This time, we'll take a look at two useful and common components. Coils are simply turns of wire wrapped around a form, while transformers change (transform) a voltage. What could be simpler?

Can we get by without coils and transformers in our Amateur Radio pastime? If we worked only with logic circuits and audio amplifiers, the answer might be "yes." But receivers, transmitters, antenna systems and most power supplies require some type of transformer and/or coil. Let's look at how coils and transformers fit into the overall scheme of things.

Meet the Coil

A fancier name for a coil is *inductor*. Each coil, depending on its diameter and the number of conductor turns it uses, has a property known as inductance. Inductance is defined as the "property of an electric circuit by virtue of which a varying current induces an electromotive force in that circuit or in a neighboring circuit."

The basic unit of inductance is the henry, abbreviated H. Our radio math can be carried out much more conveniently if we work with small fractions of the henry, such as the millihenry (mH) or microhenry (μH). A mH is 1/1000 of a henry, or 10^{-3} henry. A μH is 10^{-6} henry and a nanohenry (nH) is 10^{-9} henry. Inductance values of 1 H or greater are common only in audio and power-supply circuits. It is important to familiarize yourself with these various expressions of the henry, since you will encounter them often.

Types of Coils

Most of the large coils are wound on insulating cylindrical forms. Some are self-supporting, or "air wound." Generally, the conductor is large-diameter copper wire, but some very large coils are fashioned from copper tubing. Large conductors are needed to create a self-supporting coil. Other large coils are semi air wound; that is, they have high-grade insulating material in the form of ribs that are spaced 90 degrees apart, parallel to the axis of the coil. The coil turns are essentially air wound between the four ribs (see Fig. 1). Two

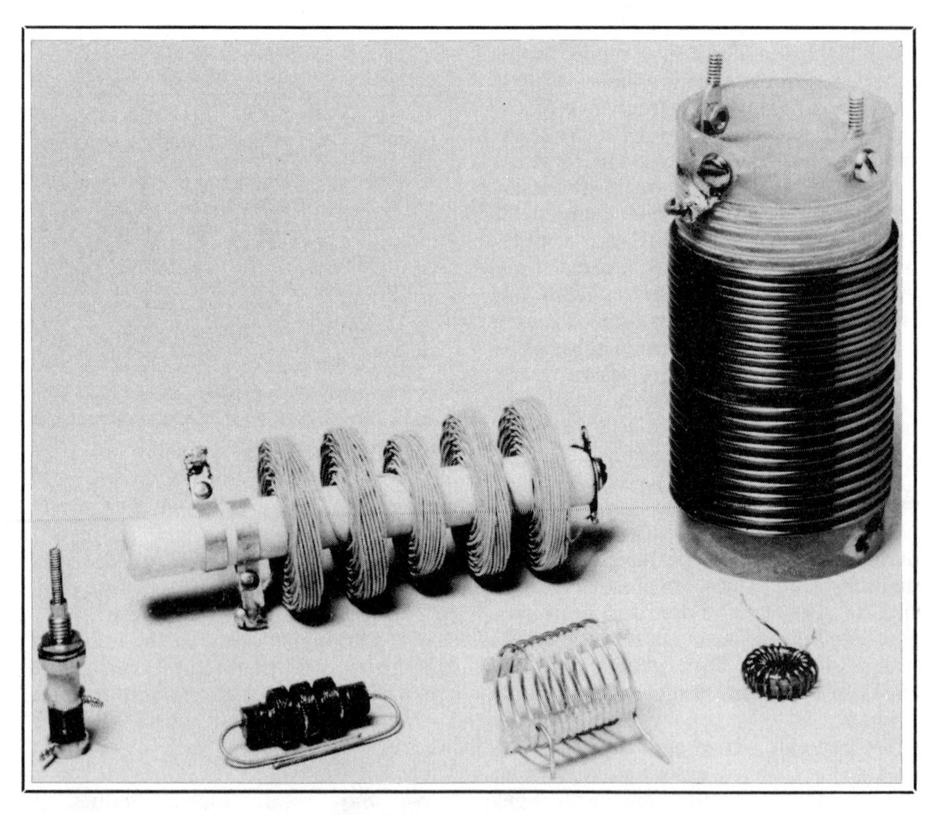

Fig. 1 — A variety of coils. Clockwise, from the left, are: a slug-tuned coil with the adjustment screw visible at its top, a high-power RF choke for transmitters, a homemade coil wound on a cylindrical insulating form, a small toroidal coil, an air-wound Miniductor coil and a small RF choke.

firms make coils of this type (Barker & Williamson Miniductors® and Poly Coils Co.).

Most of the smaller coils we will use are wound on some type of insulating form, and the wire gauge is small — usually no. 20 to, say, no. 40 gauge. Small coils are suitable in low-power circuits, but large air-wound inductors are the rule when working with high power. Today's miniature coils are wound on high-quality plastic, ceramic or phenolic forms. The coils may have only a single layer (solenoidal) of wire, or many layers may be stacked atop one another to obtain high values of inductance. The wire used in these little coils must be insulated to prevent the turns from shorting to the adjacent ones. Most large air-wound coils use bare wire for the conductors.

Another common style of inductor is the toroid. The coil is wound on a toroidal core, which is doughnut-shaped. Fig. 1 shows such a coil. The toroidal core may be made from ferrite or powdered-iron material. The exact nature of the particular core (there are many types) will determine the final inductance value for a given number of turns. This magnetic core material will always yield a higher-inductance coil than we would obtain when using an equal number of turns on a standard insulating form, or if our coil happened to be an air-wound type.

Similarly, many small coils contain a movable iron or ferrite core (slug). The slug provides a range of inductance for a specified number of turns of wire. These slug-tuned inductors are very convenient when we need to adjust the inductance for a critical value in our circuits. You will often hear an amateur say that he or she "tweaked" a circuit for correct performance. Generally, this means that the ham adjusted the slug in a coil, or perhaps adjusted a trimmer capacitor.

Some adjustable coils contain brass slugs. These are used chiefly at very high frequencies (VHF). The brass core has the opposite effect of powdered iron or ferrite: it *decreases* the inductance of a coil.

We should be aware that there is also a style of coil contained within an enclosure made from ferrite material. The coil is wound on an insulating form or bobbin, and the halves of the core material are bolted together (or cemented) over the bobbin. These units are called *cup cores* or *pot cores* (see Fig. 2). The core halves increase the coil inductance, just as an iron or ferrite slug does in a slug-tuned coil. The advantage of the pot-core inductor (or transformer) is that the outer shell provides a shield, just as would be true if a plain coil was mounted within a metal enclosure. The shielding is helpful when we want to isolate our coil from adjacent circuit elements.

No matter what form a single inductor has, it is a coil. You will hear about radio-frequency (RF) chokes. They are simply coils used for a specific purpose. You may also hear of coils being called *reactors*. In essence, these terms indicate that we're using the "same players in different games."

Fig. 2 — A pot-core or cup-core assembly. To the left is a break-down view of the core halves and the insulating bobbin that contains the coil winding. The unit at the right is a completed pot-core coil with the core halves bolted together.

Some Common Coil Applications

First, let's look at the coil symbols we are going to find in schematic diagrams. Memorize these, for you will be using them many times. The common designations are given in Fig. 3.

Thus far we've talked a lot about coils, but haven't shown examples of their use. Let's contrive an imaginary circuit for the purpose of illustration. Fig. 4 shows a suitable example in schematic form. Here we have a two-stage transistorized code transmitter. Q1 is the oscillator, and it creates our signal when the telegraph key is closed at J1. Y1 is the quartz crystal that determines our transmitter frequency. In the collector circuit of Q1, we find an RF choke (a coil) labeled RFC1. All coils are for use in alternating-current (ac) circuits: Remember that radio-frequency (RF) energy is also a form of ac. There is no such thing as a direct-current (dc) transformer. So, RFC1 is used in Fig. 4 to permit the flow of dc to the collector of Q1 while preventing, or choking, the flow of RF energy back into the +12-V voltage line. The RF choke has a value of 1 mH.

If we look to the right in Fig. 4, we will note another coil, L1. It is used to tune the output of the crystal-oscillator stage to the frequency of the crystal — 3.7 MHz. C1 is used with L1 to achieve this requirement. When the combination of C1 and L1 is tuned to 3.7 MHz, we have what is known as a *resonant* circuit, or we might say the circuit is tuned to *resonance*. We can see

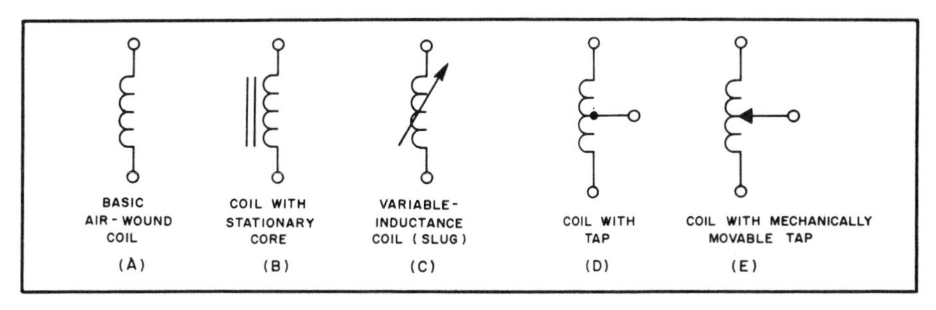

Fig. 3 — Various common symbols for coils. The example at E is for a coil with manual taps such as terminals and clip leads, or a coil with a movable contact, such as a roller inductor has.

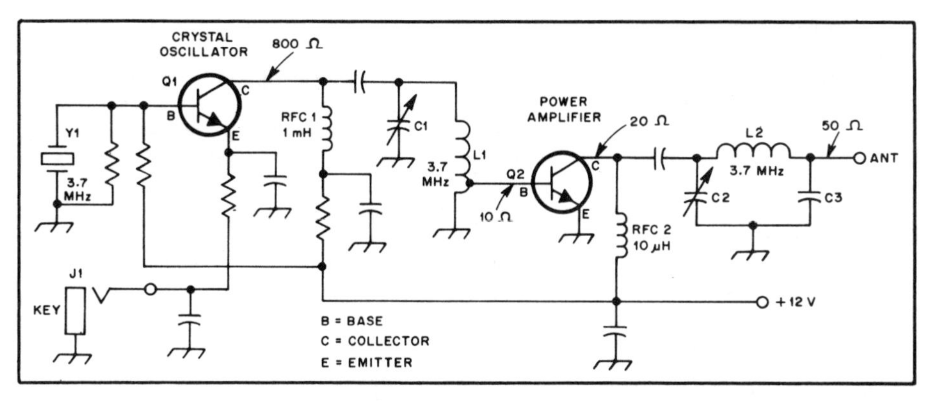

Fig. 4 — Circuit example of a transmitter that uses four styles of coil. See text for details.

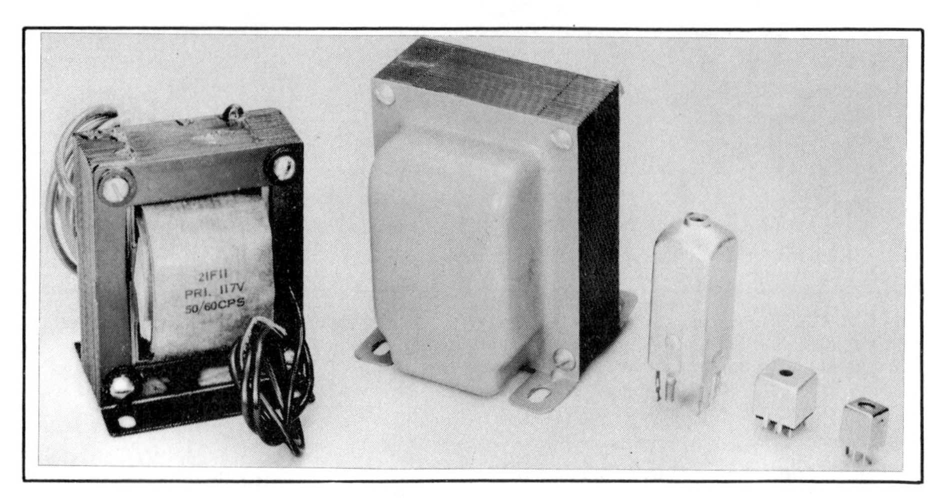

Fig. 5 — Various transformers. The one at the left is called an "open frame" or unshielded type. Next is a shielded power transformer, which is enclosed in a metal shell or case. To the right of this transformer are three styles of shielded RF transformers for use in transmitters or receivers.

that L1 serves a different purpose than does RFC1, but both are coils. L1 must have a specific value of inductance, and C1 must be set for a particular capacitance value in order to tune the circuit to 3.7 MHz. Not just any coil and capacitor combination can provide the desired resonance.

Moving to the right of our diagram once more, observe the placement of RFC2. It functions in a like manner to RFC1 — keeping the RF energy where we want it to be while permitting dc to reach the collector of Q2. In this example, we have an RF choke in the microhenry (μH) range. The reason for RFC2 having much less inductance than RFC1 is not important now. Later in our beginner's course we will learn more about such matters. You will recall that 10 μH is 1/100th of a millihenry (mH).

At the far right in Fig. 4 we have L2. It is a coil also, but in this application it serves two purposes: It is tuned to resonance by means of C2 while acting as an impedance-matching network. Our circuit example shows that the collector of Q2 looks like 20 ohms to the circuit that follows it. But, the antenna presents a 50-ohm impedance. If we are to have maximum power transfer from Q2 to the antenna, we must match the impedances of the two devices. By selecting the proper values for C2, C3 and L2 we can reach this goal.

So, we have seen three important uses for coils in Fig. 4. I should mention also that the tap on L1 (near ground) is selected to provide another impedance match. This time, we are matching the 800-ohm collector of Q1 to the 10-ohm base of Q2. The coil, L1, actually functions as a transformer under such a condition. The impedances presented by the various elements of a transistor are determined for the most part by the operating voltage and current common to the transistor. The values listed in Fig. 4 are by no means specific.

Enter the World of Transformers

From a physical point of view, a transformer is simply two or more coils wound on a magnetic core. The word "transformer" means the component can be used to transform one ac voltage to another (higher or lower than the source voltage). It also is used to transform one impedance to another, or to match unlike impedances.

A specific definition of a transformer is "a device consisting of a winding with tap or taps, or two or more coupled windings, with or without a magnetic core, for introducing mutual coupling between electric circuits." Transformers that have no magnetic core material are used at radio frequencies, but many RF transformers do contain core material. Conversely, coreless transformers are not suitable for use at audio frequencies and lower. Fig. 5 shows a variety of transformers as assembled units. The larger the size, the greater the power-handling ability of the device.

Transformer Applications

I'm sure you are aware of the large transformers found on utility-company poles throughout your area. These "pole pigs," as some amateurs call them, are used to reduce the potential on the power line before it is routed to the consumer. The power lines that crisscross the country carry thousands of volts. It would be unsafe and impractical to route so high a potential into our homes. Therefore, the existing power-line voltage is dropped to 234 V for entry into our homes.

You will also find power transformers in your TV set, hi-fi gear and ham radio equipment. These are used in the equipment power supplies to change the 117-V ac-line level to some higher or lower voltage. The voltage chosen depends on the requirements of your equipment. After the voltage is lowered or raised by the transformer, it is converted to dc voltage by means of *rectifiers* (usually semiconductor diodes). Then, the not-so-pure dc voltage is filtered to remove any ac energy that may still be present after rectification.

Various types of transformers are shown schematically in Fig. 6. Illustration A shows the basic arrangement for a transformer that has two windings — a primary and a secondary, as we call them. The two parallel lines between the windings signify that a magnetic core exists. It might be made of iron, powdered iron or ferrite material, depending on the application. Voltage is specified in Fig. 6 as E, and it can be of any frequency in the ac range. The proper core material must be used for the frequency of operation if the transformer is to function correctly, however.

Next, let's consider the transformer of Fig. 6B. It is similar to the one shown at A, except that it steps down the voltage we might apply to the primary winding. The ratio of the turns of the windings determines what the transformer output voltage will be. The smaller the number of secondary turns, the lower the output voltage.

Fig. 6C shows a transformer with a number of taps on the secondary winding. Under this arrangement, we may have a variety of secondary voltages available. The location of the tap, respective to the number of turns for both windings, will determine the output voltage. Fig. 6E shows a unit that can achieve the same results, except that separate windings are used to obtain the different output voltages.

At Fig. 6D we have a tuned transformer. This is a common type that we will encounter in working with RF circuits. Because the transformer primary winding and capacitor C form a resonant circuit at a desired frequency, we are actually dealing with what is called a *narrow-band* transformer. Untuned transformers respond to a broad range of frequencies, so they are known as *broadband transformers.* The core material in the transformer of Fig. 6D is adjustable within the coil winding. This slug enables us to tune the transformer precisely to the operating frequency. For RF work, the core will be made of powdered iron or ferrite.

Finally, we see an audio transformer at Fig. 6F. It is similar to the transformers shown at A and B, except that we have a center tap in the primary winding. This allows us to provide what is known as "push-pull" operation for the two output tubes or transistors in the audio amplifier. In other words, we will achieve a desired balanced condition for the amplifier devices.

Audio transformers are used also to ensure an impedance match between the amplifier output and the load, which in our example is an 8-ohm speaker. The impedance transformation is related to the turns ratio of the windings. It is the square of the turns ratio. Hence, a turns ratio of 3:1 will yield an impedance ratio of 9:1. Conversely, a 12:1 impedance ratio will be

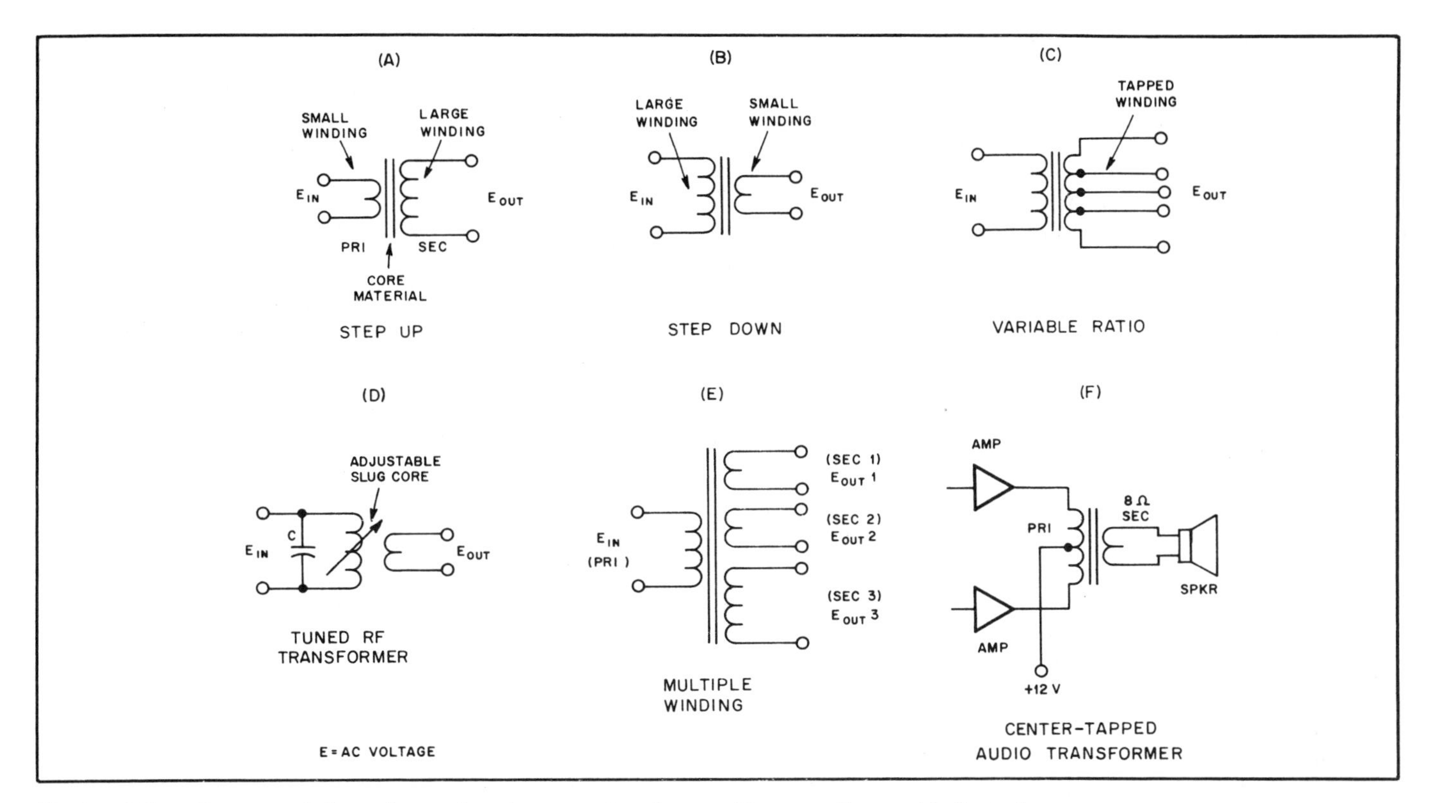

Fig. 6 — Schematic representations of a number of common transformers. These are discussed in the text.

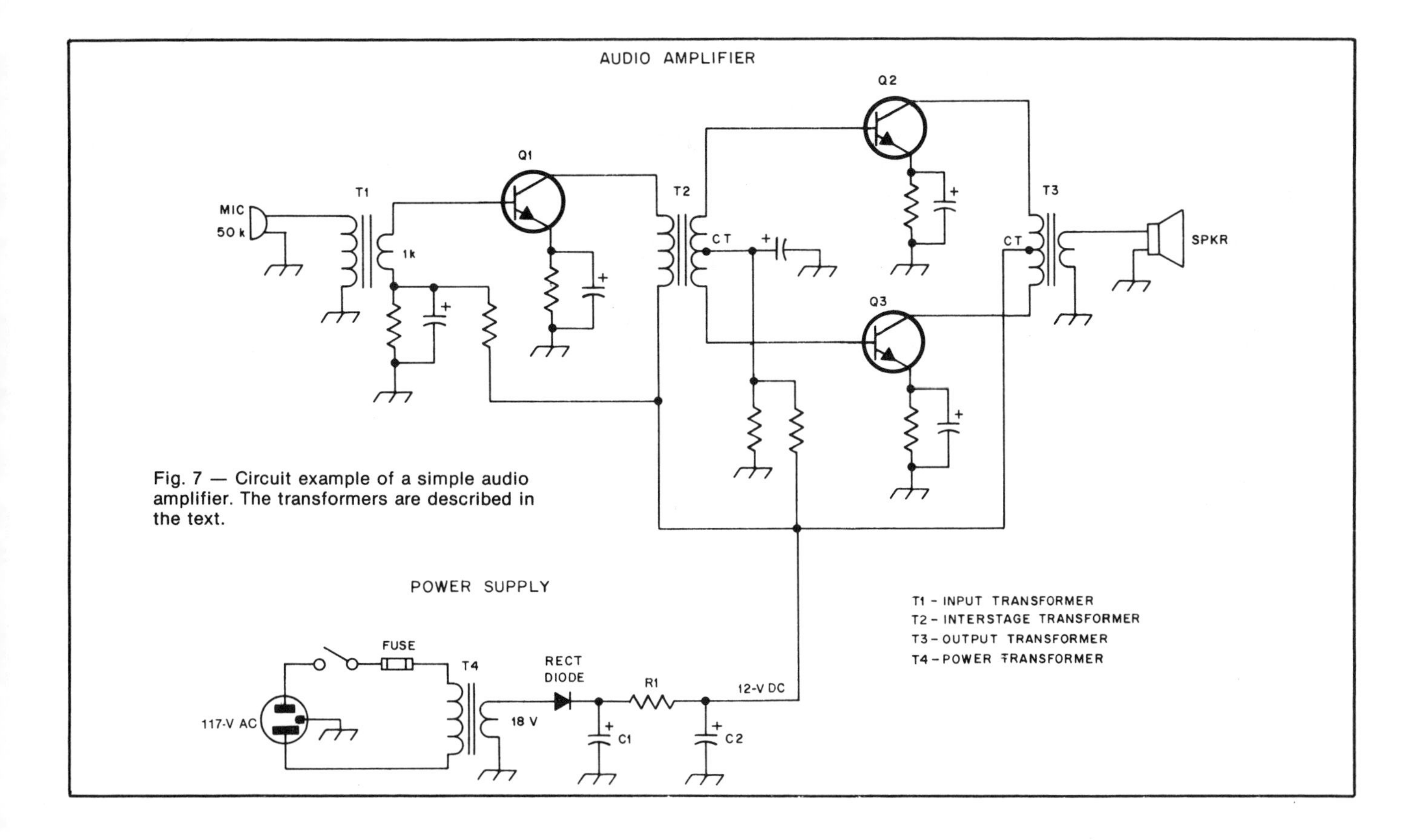

Fig. 7 — Circuit example of a simple audio amplifier. The transformers are described in the text.

had when the turns ratio is 3.46:1. The transformer voltage ratio, however, is the same as the turns ratio. Memorize these facts for later use.

Fig. 7 shows a hypothetical audio-amplifier circuit in which some transformers are used. You will note that we have a power supply in our circuit. It also uses a transformer, T4. It steps the voltage down from 117 to a more manageable 18. The diode rectifier converts the ac voltage to dc voltage. The remaining unwanted ac energy is filtered out of the +12-V line by means of C1, C2 and R1. T4 also isolates us from the 117-V wall outlet, helping to prevent shock hazards.

Transformer T1 is used to match the high impedance of our microphone to the low impedance of the Q1 transistor input. So, we can think of T1 as a matching transformer, or an input matching transformer. T2, on the other hand, is an interstage transformer with a center-tapped secondary winding. The split winding enables us to supply audio energy in push-pull (balanced) to the push-pull output transistors. It can be used also to match the output impedance of Q1 to the input impedance of Q2 and Q3 if the proper turns ratio is chosen.

The output transformer, T3, functions as does the example of Fig. 6F, which we have already discussed. We have not assigned parts values to any of the circuit components, since this is purely an imaginary circuit. In reality, most modern audio amplifiers that use transistors do not employ audio transformers, but they were standard fare in the vacuum-tube days and during the early days of transistors.

Glossary

henry — basic unit of inductance. Abbreviations are H: henry; mH: millihenry; μH: microhenry; nH: nanohenry.
inductance — a property of an electric circuit by which voltage is induced in it by varying current in the circuit itself or a neighboring circuit.
inductor — a coil or transformer winding, with or without a magnetic core, for introducing inductance into an electric circuit.
magnetic core — one of various materials, such as iron, brass, powdered iron or ferrite, contained within a coil or transformer winding to increase the inductance over that which would exist with no core material. It concentrates an induced magnetic field in a transformer or coil.
potential — the relative voltage or voltage level in an electric circuit.
reactor — a device used for introducing reactance in a circuit. An inductive reactor, or inductance. Coils and transformer windings exhibit reactance, hence can be referred to as reactors.
rectifier — a device used to convert ac to dc. Semiconductor or tube types of diodes are used for this purpose.
resonance — a point at which a coil and capacitor combination are set for the same or zero reactance at a chosen frequency. A condition under which the coil and capacitor are adjusted so as to be tuned to a specific, chosen frequency. The circuit is then said to be "resonant."
shielding — the use of metal or other conductive material to prevent magnetic or capacitive coupling between circuit elements. A shield is an electrical barrier for ac energy.
signal — ac or dc current varied according to the information it carries.
quartz crystal — a mineral that resonates at a precise frequency according to how it is cut.
toroid — doughnut-shaped device, such as a toroidal core for coils and transformers.

Coil and Transformer Power Capability

The greater the power a transformer must accommodate (watts = E multiplied by I, where E is voltage and I is current), the larger the wire size and the greater the core area. The core material plays an important role in the power rating too, as some materials are more efficient than others. The large wire is needed to reduce the resistance (and heating) of the windings. Also, the greater the winding resistance, the higher the transformer losses. An ideal transformer would be cool to the touch after many hours of operation, but this is seldom the case. Most transformers in power supplies are warm or quite warm to the touch after they have been on for a period of time. This heat causes wasted power and reduced efficiency.

Coils that must handle RF power also can become warm. To reduce resistive losses, it is wise to use large-diameter wire for such coils. High-quality insulation should always be used in coils and transformers to prevent arcing between the windings, and to minimize losses.

Let's Summarize

What have we learned? First, that coils can take many shapes. They can be built for fixed values of inductance, or they can be made variable by using a movable slug inside them. They are used in all manner of radio circuits and at many power levels. Coils are also known as inductors, and they may be wound on magnetic cores or can be built as "air-wound" units with no core.

Transformers are used from the power-line frequencies (50 or 60 Hz), through audio frequencies, and into the high RF range. They can be narrow band or broadband, and they may also have cores or no cores. They are used not only to step up or step down a specific voltage level, but may serve as impedance-matching devices between components of unlike characteristics. The impedance ratio of transformers is the square of the voltage or turns ratio and vice versa.

Coils and transformers are among the common radio parts we will be working with during our amateur careers. Detailed information about them can be found in *The ARRL Handbook* and other ARRL books.

The World of Switches and Relays

Part 6: Mechanical and electronic switches are important parts of most radio circuits. Let's examine mechanical switching devices and become acquainted with how they look and perform.

What could be more ordinary than a switch? After all, we have them on our appliances, on the walls in our homes and on the instrument panels of our automobiles. Switches are as old as electricity, and they come in many different shapes and sizes. For that reason, we must know how to select the proper switch for each application.

There are many things to consider when selecting even a simple, inexpensive switch. Among them are insulation quality, mechanical durability, the number of switch terminals (poles and contacts), current-carrying ability of the electrical contacts, and physical size. We can add to this list the cost of the switch versus the well being of our hobby budget! In other words, "any old switch is not necessarily the right one for a specific job."

We need to understand the circuit requirements and choose a suitable switch for that circuit. The same is true of relays, which are electrically operated switches. They differ from ordinary switches by virtue of being actuated or energized remotely by a mechanical or electronic switch. In effect, we have a mechanical switch being used to turn on a remote switch.

In some instances, an event that takes place within an electronics circuit will cause the relay to switch on or off automatically. The TR (transmit-receive) relays in our ham radio transceivers are examples of devices that are controlled by the circuit within the transceiver. Some are called VOX (voice-operated) relays because they actuate when we speak into the microphone. When we stop talking they disengage, thereby connecting the antenna and certain operating voltages to the receiver circuit. When the

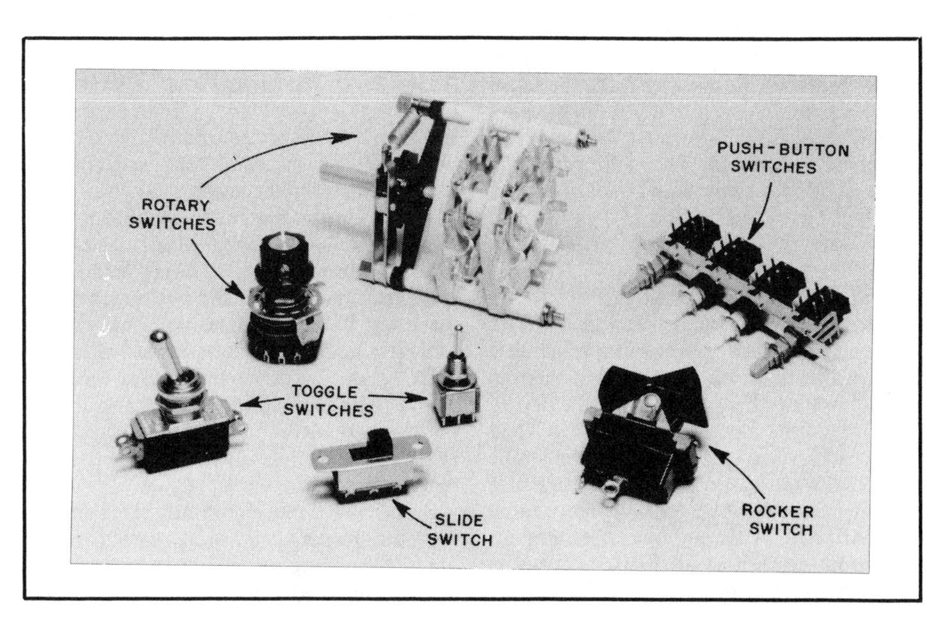

Fig. 1 — Various types of switches.

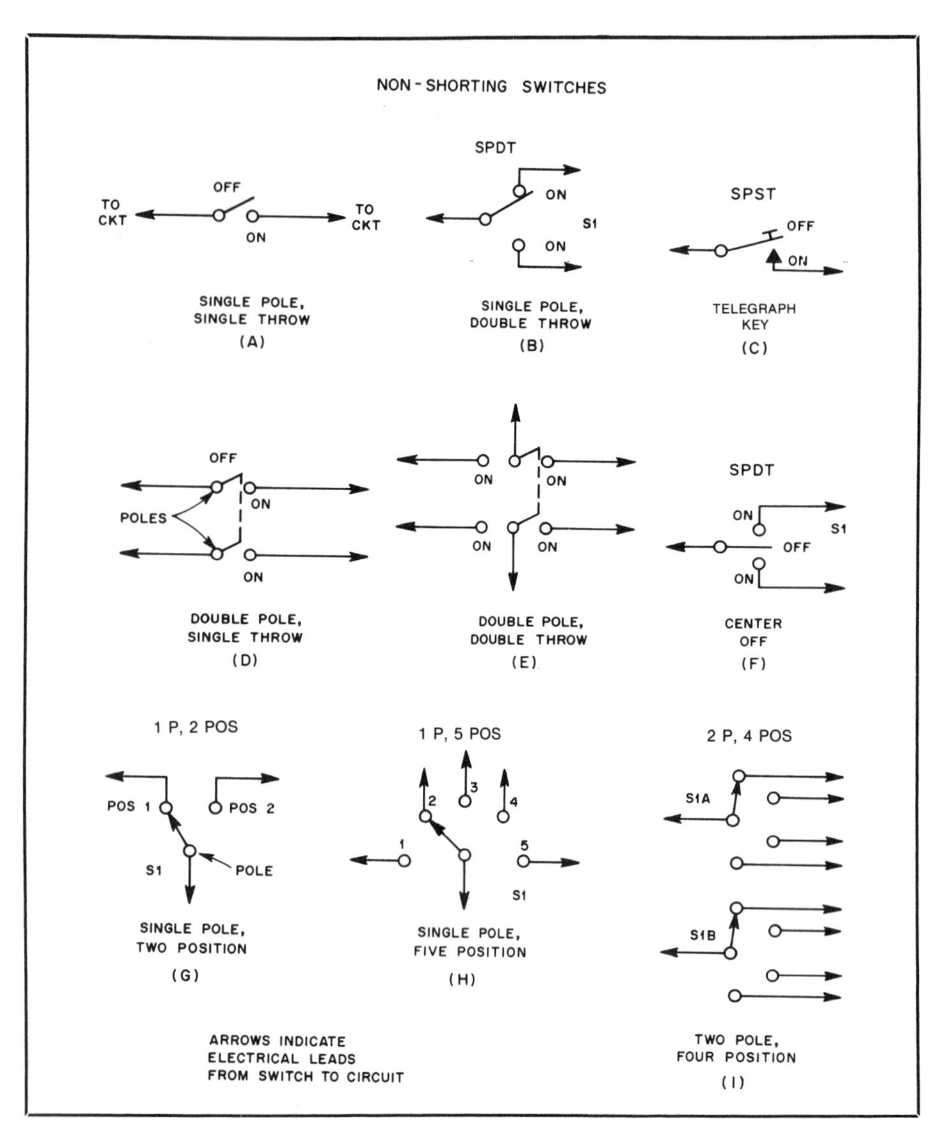

Fig. 2 — Electrical symbols for a number of common switches used in Amateur Radio work.

relay is activated by our voices, the antenna is switched to the transmitter, along with specific operating voltages. Relays are often used to control an antenna function at some point that is a distance from the radio room. Generally, such remote relays are actuated within the ham shack by means of a mechanical switch. There are countless applications for relays and special switches that can be turned on and off from a distant point.

Switches and Their Circuit Symbols

Each style of switch is represented by a different electrical symbol in a schematic diagram. It is important that you become familiar with the symbols if you are to understand the function of the switch when you read a diagram or draw one of your own. I'd like to suggest that you practice drawing these symbols until you have them memorized. Although you will find numerous variations of the basic symbols for switches in amateur magazines, they follow a pattern that should be easy to recognize. The ARRL employs the established standards for electrical symbols, as used by the industry and the IEEE (Institute of Electrical and Electronics Engineers). As we learned in an earlier installment of this series, many of the other publishers use nonstandard symbols to create a distinctive "style."

Fig. 1 shows an assortment of switches. We can tell from the illustration that switches come in many shapes and sizes. Generally, the smaller the switch assembly the lower its power-handling ability (current and voltage rating). Most reputable switch manufacturers can provide the consumer with maximum safe ratings for voltage and current. Some manufacturers imprint the ratings on the switch for our convenience.

Some of the more commonplace switch symbols are depicted in schematic form in Fig. 2. The simplest switch is shown at A and C. This is what we call an SPST (single pole, single throw) switch. A telegraph key (C) is actually an SPST switch that we open and close by hand to form the Morse code characters. When we use the switch at A, we can complete only one circuit. This type of switch is common in our TV sets, lamps and household appliances. But, if we want to control two circuits, we must use an SPDT (single-pole, double-throw) switch (Fig. 2B). Here we have two ON positions. As we examine the progression of switch symbols in Fig. 2, we will recognize the utility of switches with more than one pole and two throws.

The switches from A through F in Fig. 2 are considered to be *toggle* switches. This style of switch has a bat handle or flat lever used to operate the switch. These switches are available also as "rocker switches." This variety is built with a plastic bar that is pressed on one end to operate the switch to close one circuit; the opposite end of the rocker bar is depressed to change to another circuit. There is a rocking action to the bar; hence the name "rocker."

Switches can be obtained also with a "center off" position. We can see this at F of Fig. 2. Here we have two ON positions and a center position that is OFF. In the off condition our switch pole is in midair, so to speak.

The switches shown schematically from G through I of Fig. 2 are what we refer to as rotary switches. This means that instead of an up-down or side-to-side action (as with toggle switches), we rotate the switch shaft and pole clockwise or counterclockwise. The contacts that the pole conductor touches during rotation are mounted on a thin wafer of phenolic, plastic, ceramic or steatite material. The usual name for these switches is "rotary wafer switch." Rotary switches may contain several decks of wafers, and may have 30 or more positions. These are often called multiwafer or multigang switches.

Some Other Switches

The rotary switch may take another electrical form, as we can observe by looking at Fig. 3. This is known as a shorting type

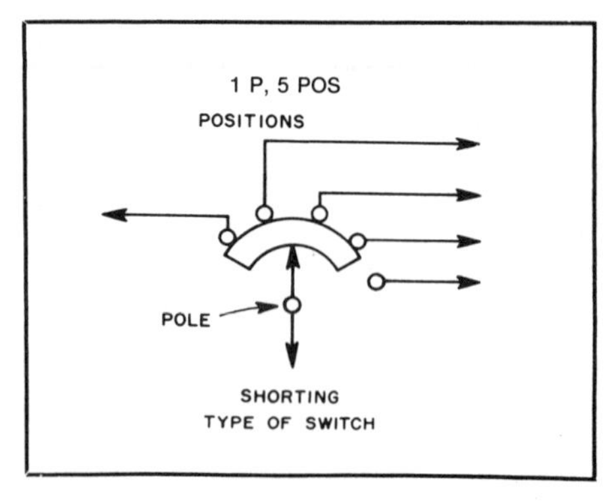

Fig. 3 — One style of shorting switch (see text).

of switch, (the switches of Fig. 2 are nonshorting units). How do these switches differ? Well, if for some reason we want to short out the circuits that are connected to the switch terminals while only one event is controlled by the nonshorted contact, we may use a shorting style of switch. Perhaps a more common type of shorting switch is the rotary type with a pole contactor that is sufficiently wide to cause any two adjacent wafer contacts to be shorted together as the switch is cycled to a new position. Once the switch is in the intended position, the shorted condition ceases. This wiping action is useful when we do not want to interrupt one circuit until another is closed. These switches are not used very often, but we should understand how they can be used in case a special application comes up.

We may hear about *mercury switches* in our discussions with other hobbyists and amateurs. This variety of switch is operated by changing its physical position. It contains two contacts that are open when the assembly is in one position. When the position is reversed, the mercury within the switch changes position and acts as a conductor across the internal contacts. These motion switches are used in automation and burglar-alarm systems, to name only two uses.

Rotary switches are available also for remote actuation by means of electrical current. They are sometimes called "solenoid switches." They are standard rotary switches that have a coil to which an operating voltage can be applied from a distant point. Each time a pulse of current is sent to the switch coil, the pole of the switch will move to a new position. Another version of this switch is the "stepping switch." It contains rows of contacts to which we may attach wires that go to various circuit points. Each time a pulse of current is supplied to the coil on the switch, the pole(s) moves to a new position. Telephone companies used stepping switches in their control circuits for many years.

You will also hear about an interesting gadget called a "proximity switch." It is merely a mechanical or electronic switch that actuates when a conductive object is placed near it. To operate in this manner, it must have some allied electronics to control it. Switches of this variety can be actuated also by changes in light, moisture or heat levels.

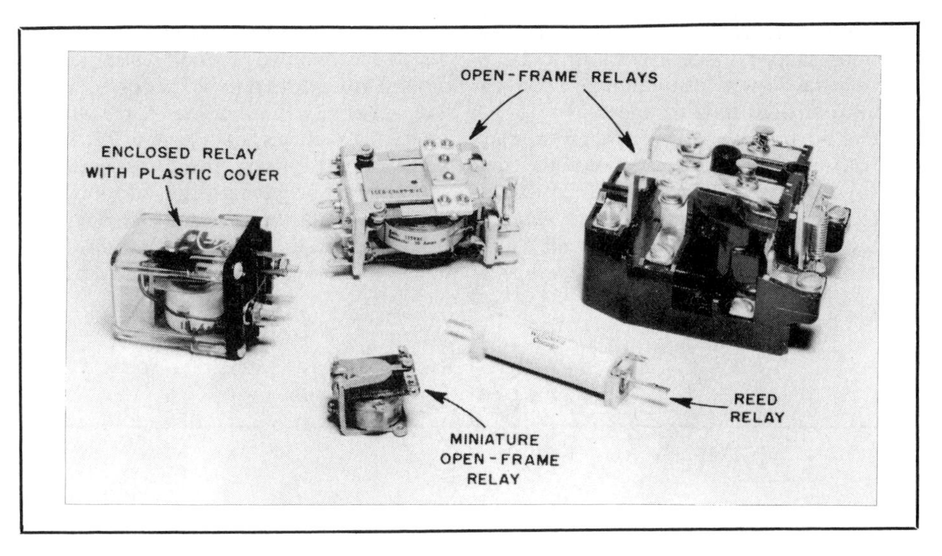

Fig. 4 — Various forms of relays.

Some Aspects of Relays

It is easy to compare a relay to a switch, for in fact it is a switch of sorts. That is, a relay switches circuit connections in accordance with the commands given to it. These commands may be at the direction of a person, or a sensing circuit may respond to some event (such as heat, voltage or current changes) and in turn cause the relay contacts to open or close. For a relay to operate, it must have a certain ac or dc current flowing through the field coil. A field coil is a multilayer winding of insulated copper wire that is contained on an iron-core bobbin. When the electrical field of the coil becomes intense enough, it magnetizes the iron core and pulls the relay pole (arm) down. This action closes the electrical switch contacts of the relay. When the current flow is removed from the relay field coil, the magnetism ceases and the relay pole returns to the de-energized position.

Relays may have several poles, as do mechanical switches. They can be compared to toggle switches in their capability because, like the toggle switch, a relay is a two-position device. This is not true, of course, with rotary switches.

Ac-operated relays have a copper half-round plate at the top of the field coil. It is positioned to "shade" the field in such a way as to prevent the relay from buzzing when it is activated. This copper plate is known as a "shading pole." The cause of the mechanical vibration of an ac relay is the pulsating 60-Hz current from the wall outlet. No such problem is found when using dc relays. Sometimes it is necessary to loosen the field-coil assembly and rotate it in an effort to find a position where annoying "chatter" will not occur.

It would be impractical to describe every type of relay that we can obtain. We can generalize by saying that the major differences in relays are the operating voltage, current rating of the contacts, size of the unit, mechanical characteristics and mounting style. Some relays have plastic covers to help keep dust and moisture away from the electrical contacts. Others have no enclosure; they are called "open-frame relays."

We must pay attention also to the type of insulating material that isolates the relay contacts from one another. This material must not break down and burn in the presence of the voltage we apply to the relay contacts. The size of the contacts is important also, for they must be able to pass the current that flows through them. If the contacts are too small, they will become hot and may even melt or become charred. In less-severe cases, the contacts can become pitted and covered with oxide. This makes them incapable of making a proper electrical connection. Under some conditions of pitting the contacts may weld together, which will prevent the relay from operating normally. We can clean dirty relay contacts with a piece of emery cloth or a fine file.

There is a style of relay that has its pole arm *inside* a field coil. When the field is turned on, the pole moves into the contact state, closing the circuit. These devices are known as "reed relays." They are capable of cycling at a fast rate with little "contact bounce." This condition often occurs when large relays are used for fast switching. The contacts bounce apart and cause circuit-performance problems (jitter). Some amateurs use "bug" CW keys that mechanically send automatic dots. The dashes are made in the usual matter — by pressing the dash paddle. Some hams have been known to stuff a small piece of foam plastic or a filter tip from a cigarette into the loop-shaped dot-contact spring to damp the inertia. This helps to minimize contact bounce at high sending speeds.

The operating voltage for a given relay coil depends on the dc resistance of the coil. In order to set up a magnetic field that is sufficient to close a relay, a specific amount of current must flow through the coil. Since current is determined by E/R (where E is the operating voltage, and R is the coil resistance in ohms), the coil resistance and applied voltage need to be the correct value. Relays are manufactured for standard operating voltages, such as 5, 6, 12, 18, 24 and 48. It is necessary, then, to select a wire size and winding length that produces the desired resistance to obtain the current needed through the field winding. Each gauge of wire has a specific dc resistance (in ohms/ft). Hence, it is easy to make a field coil to match the operating voltage and pull-in characteristics of the relay.

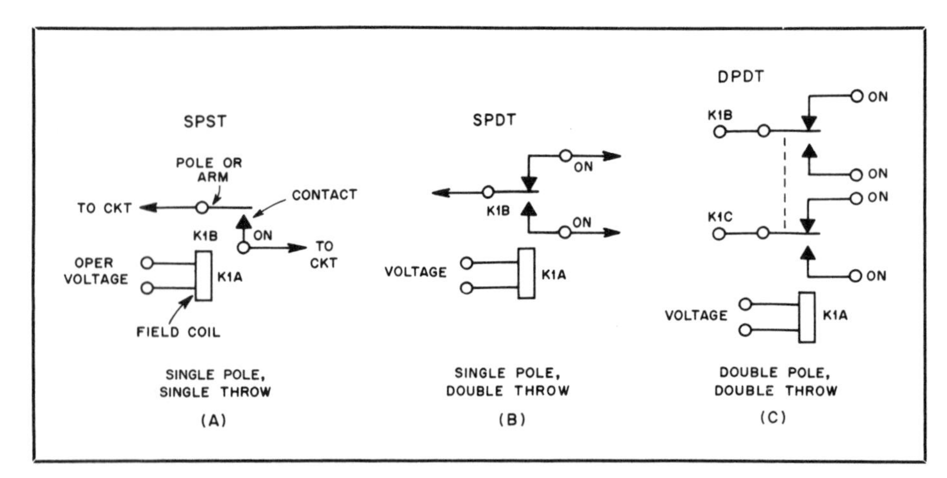

Fig. 5 — Electrical symbols for some common relays.

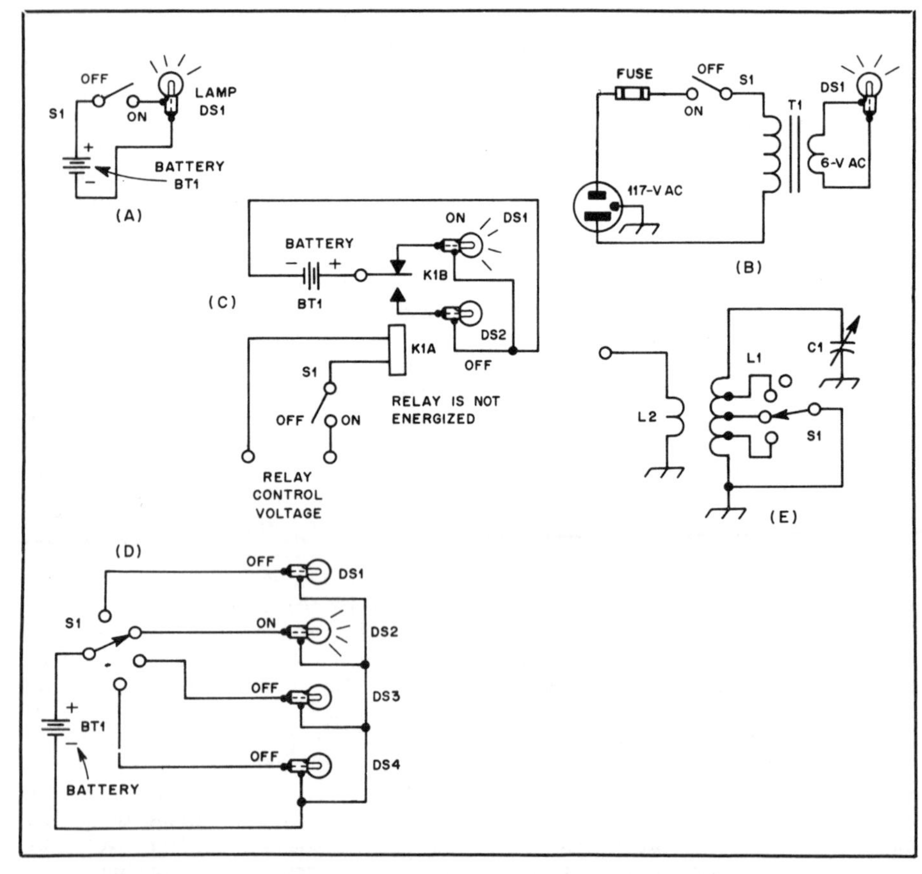

Fig. 6 — Some practical examples of how to use switches and relays. See text.

Glossary

bobbin — an insulating coil form upon which magnet wire is wound before it is added to a relay or other electrical mechanism.
contact bounce — a condition that can occur when a relay is cycled, causing the electrical contacts to bounce apart momentarily. This causes the circuit being switched to be interrupted during the period of the bounce.
electronic switch — an electrical rather than mechanical switching circuit. Generally, such switches consist of diodes, transistors or ICs.
pole — the movable portion of a switch or relay that makes electrical contact with the remaining conductors of a switch or relay.
proximity switch — a type of switch that is actuated by the nearby presence of a conducting object or magnet.
reed relay — a type of relay that has the pole piece contained inside the field coil. The pole arm is like a small, thin reed.
relay — a mechanical switching device that is actuated when sufficient current flows through its field coil, thereby causing the field-winding core to become magnetized. This action moves the pole arm to make and/or break contact with a circuit.
rotary switch — a switch that has a shaft that causes the switch pole to come in contact with the mating electrical contacts on the switch wafers. A rotation of the shaft is needed to operate the switch.
shading pole — a copper, half-round plate that is affixed to the top of a relay field coil when the relay is intended for ac operation. The shading pole prevents the relay from buzzing when it is energized.
solenoid switch — a switch that can be operated remotely by means of a field coil to which ac or dc voltage is applied.
stepping switch — similar to a solenoid switch, but with contacts arranged in a row (part of an arc) rather than on a circuit wafer.
VOX — voice-operated relay and control circuit used in a transmitter or transceiver.

Fig. 4 shows an assortment of relays. The symbols for a few relays are provided in Fig. 5. Please note that when multiple-section switches or relays are used they are labeled in accordance with their sections. For example, the relay in Fig. 5C has three parts: the field coil (K1A), one set of contacts (K1B) and a second set of contacts (K1C). This type of labeling permits us to place sections of the switches or relays in different parts of a diagram without losing track of which part belongs with another. The parts may be spread about in a diagram for the purpose of minimizing connecting circuit lines and crossovers. This helps to "unclutter" a diagram and make it easier to follow.

Some Practical Examples

Let's see just how we might use switches

and relays, singly or in combination, to make certain things occur. Fig. 6 contains some simple circuits to demonstrate what can be done with switching arrangements. Circuit A shows how a lamp can be turned on and off with an SPST switch — the kind we will find in most table lamps and wall panels. Fig. 6B shows a lamp being operated from a 6-V ac transformer. We have reduced the ac line voltage from 117 to 6 by means of a step-down transformer, T1. This is necessary to obtain the right operating voltage for DS1, a 6-V lamp.

Fig. 6C shows how we can use a relay to control two lamps, DS1 and DS2. Although this circuit is somewhat absurd and would not normally be used, it does illustrate how a relay can be used to provide remote control of a circuit. When the relay is energized (S1 switched to ON), we will see DS2 illuminate and DS1 will no longer glow.

At Fig. 6D, we find a rotary switch being used to control four lamps. DS2 is illuminated in the example, because the switch pole and related contact are permitting current to flow through only DS2. The lamps will light in sequence as we rotate S1.

An example of coil-tap selection is shown in Fig. 6E. As S1 is rotated clockwise more and more of the coil is shorted out, decreasing the effective inductance of the tuned circuit. This is a common arrangement in transmitters and receivers when the operation is changed from one band to another. Similarly, the switch could be used to select individual coils for the various frequencies of interest.

Potential Problems

We learned earlier that the insulation on switches and relays must be suitable for the application we have in mind. We can adopt a role of thumb concerning this matter: For operating voltages (dc, ac or RF), we can do rather well with plastic or phenolic insulating material if the voltage is low (less than a couple of hundred volts) and if the RF power is under, say, 25 W. For high voltage and high RF power levels, ceramic or steatite insulation is recommended. It will sustain high voltage without breakdown, as compared to phenolic and plastic.

Relays with gold-plated contacts cost more, but they are less likely to arc and become pitted. Furthermore, they will not become contaminated by oxidation. Determine the current that your relay or switch contacts must pass, then to be safe, use a relay or switch rated in excess of the current in your circuit. I like to use a switch or relay contact rated for 2 A or more in a circuit that carries 1 A.

Tag Ends

I hope this article has proven informative in your quest for basic data about electronics. I'd like to suggest that you obtain some switches and relays, a few bulbs and some batteries. Spend a couple of hours hooking these devices up in order to see what can be done with them. You might even amaze your friends with an array of flashing lamps! Good luck.

Meet the Versatile Diode

Part 7: Diode devices have been with us since the beginning of radio. But in recent years they have taken many new and sophisticated forms that make them useful for a host of applications.

Let's become acquainted with one of the most commonly used electronic devices — the diode. Diodes have long been used for changing ac to dc voltage (rectifiers), and they have been put to work for many years as *detectors* to convert radio-frequency waves to audio-frequency ones so we might listen to music and conversations that are broadcast through the ether.

Diodes were a vital part of the first radios. Those receivers consisted mainly of a galena-crystal detector (diode) with a "cat's whisker" wire that was adjusted to touch the crystal in a spot that would cause diode action. The radios also used a large coil of wire with tap points for tuning the system to the desired radio-station frequency. We knew them as "crystal sets."

Many an early-day experimenter hunted for hours with the cat's whisker knob in hand, trying to find a "hot spot" on the galena crystal. A truly good hot spot was not only hard to find, but it might never be found a second time! When it *was* located, the broadcast station would be much louder than when other detector spots on the crystal were used.

No longer must we deal with such crude methods for radio-wave detection. If diodes are used for that purpose, they come in tiny packages with wire leads, and we need only to solder them into our circuit. Many large diodes, designed to accommodate large amounts of current and voltage, have threaded studs on their cases, making them easy to mount on a heat-conducting surface for cooling purposes.

Fig. 1 shows a number of diode types that are in use today. We can see that it is not always easy to identify a diode by its outward appearance. Many resemble small resistors or capacitors, while others look like transistors. The tip-off for most diodes is that they have but two terminals, whereas other semiconductor or tube devices have three or more terminals.

Whatever type of diode we may use, it is a *passive* component. That means it doesn't require an operating voltage to make it function in a circuit. This is not to say that operating voltage is not applied to some diodes in special instances, for to make specific events take place we must at times place a voltage on the anode or cathode of a diode. More about that later.

Diode Basics

Fig. 2 shows the symbols for some of the common vacuum-tube diodes that were popular in the "tube era." Although some equipment still contains tube types of diodes, modern apparatus utilizes semiconductor (solid-state) diodes. They are more efficient and much smaller in size. You should be aware, however, of the general format for early-day diodes. The principle of operation was about the same as it is for semiconductor diodes.

Our modern diodes take the general forms illustrated in Fig. 3. Unlike the tube diodes of Fig. 2, there is no heater that must be warmed up for current to flow.

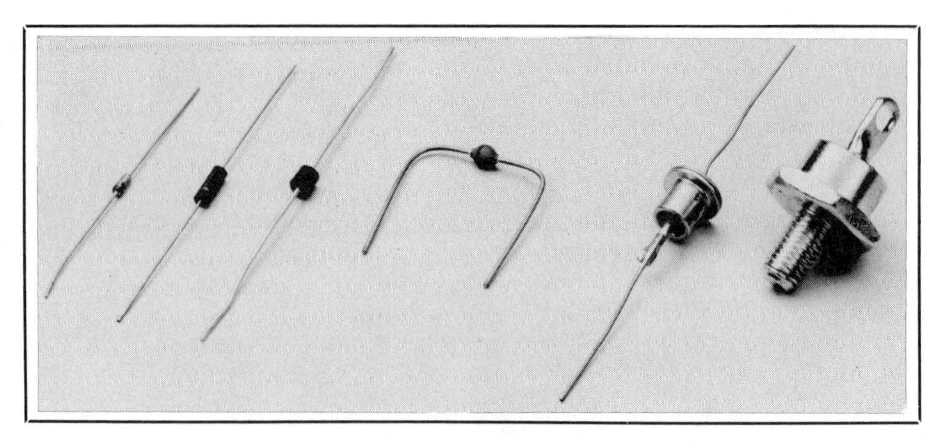

Fig. 1 — Left to right are small-signal diodes, medium-size rectifier diodes and a large stud-mount power diode. Each has one thing in common — it is a two-terminal device that can change ac into dc.

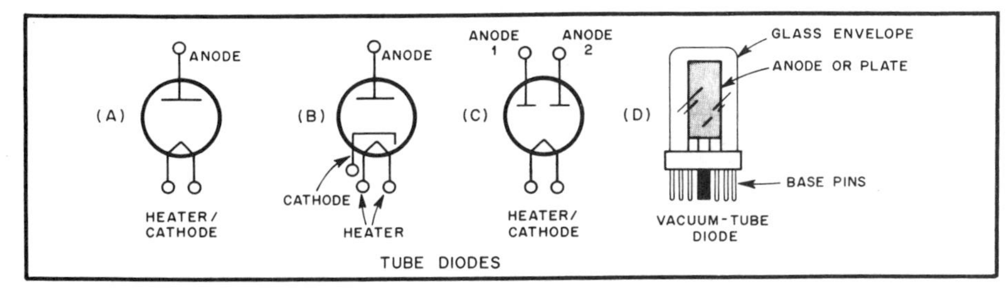

Fig. 2 — Early-day radio circuits used tube-style diodes for converting ac to dc. Although some tube diodes are still in use, they have been replaced for the most part by solid-state diodes.

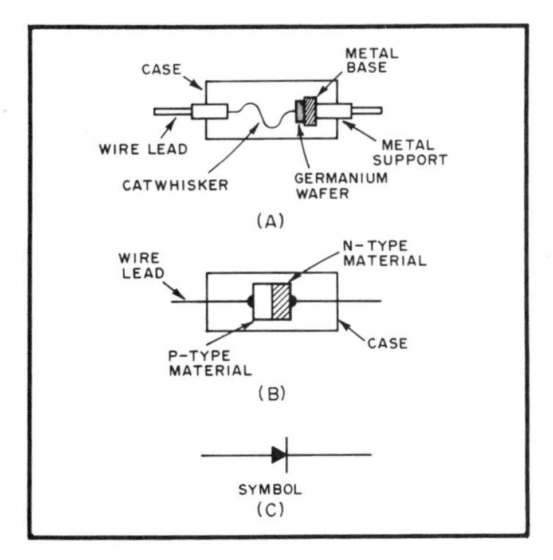

Fig. 3 — The point-contact diode is shown at A. A junction diode is at B, and the standard electrical symbol for a solid-state diode is at C.

The heater also served as the cathode in some tube diodes, while others contained a separate cathode that was warmed up by a heater. There is no warm-up time for a solid-state diode: It begins its function the moment energy is applied to it. Ironically, the galena (and carborundum) crystals of early radios were solid-state diodes, but the concept languished for many years before progress was made. The need for good detector diodes in radar systems served as the springboard for advances in the technology during WW II.

The early type of solid-state diode was the point-contact variety, as shown at Fig. 3A. A tiny wire contacts the germanium material to form the equivalent of a galena crystal and cat's whisker from days gone by. Although point-contact diodes are still available, the major usage is for junction diodes of the kind shown at Fig. 3B. Two types of semiconductor material (P and N) are formed into a sandwich to provide a rectifying junction that will permit diode action. The electrical symbol for a semiconductor diode is presented at Fig. 3C. The end with the arrowhead is the anode, and the part with the single line is the cathode.

There are two fundamental types of semiconductor diode — silicon and germanium. The essential difference is that it takes slightly more applied voltage (roughly 0.7 V) to make the silicon diode conduct and commence functioning. The majority of our modern diodes contain silicon junction material. The 1N914 small-signal diode is an example of a silicon unit, whereas the almost generic 1N34A small-signal diode is made from germanium material. You will see many circuit examples that call for those two diodes. The approximate conduction or barrier voltage of a germanium diode is 0.3 V, compared to the 0.7 V of a silicon diode. That is, the diode must have at least 0.3 or 0.7 V across it in the correct polarity before current will flow.

How Diodes are Used as Power-Supply Rectifiers

We mentioned earlier that diodes can change ac to dc. Nearly every power supply (exclusive of those that use batteries) contains diodes that serve as rectifiers for changing ac to dc. Let's look at Fig. 4 to see how the diodes might be connected in a circuit to serve our need.

Let's suppose we wanted to develop + 12 V for powering a small CW transmitter. We would have to step down the wall-outlet voltage from 117- to 12-V ac. T1 of Fig. 4 would accomplish that. But we still need to change the ac to dc. If we did not rectify the ac voltage, our transmitter signal would have a bad hum on it, caused by the 60-Hz ac wave from the wall outlet or transformer secondary winding. Similarly, we would hear a raucous hum in the speaker if we used an ac voltage to power our receiver.

So, to obtain dc output from our power supply, we will add D1 and D2 of Fig. 4A. The rectifying action of the diodes will change the ac to pulsating dc, and will double the power-line frequency to 120 Hz. Remember, it is 60 Hz to start with. The pulsating dc will still cause hum on our transmitter signal, so we have to take another step in our design. Fig. 4B shows the same circuit, but we have added two filter capacitors (C1 and C2) and a filter choke (L1). These components will smooth the pulses that otherwise could cause hum.

Notice that the dc output now has but a slight ripple. This would be so small that we might not hear it on our transmitter signal or in the speaker of a receiver. We could see the ripple if we connected a sensitive instrument (such as an oscilloscope) to the output dc line. An ideal power supply would have no ripple, and only a straight line would appear on the tube face of a 'scope. These illustrations represent the basis of all power supplies, but some use four diodes in what is called a full-wave bridge circuit. Even a single diode can be used alone to form a half-wave rectifier.

Diodes as Signal Detectors

Let's return to the subject of employing diodes to change RF energy to sound energy (detectors). A diode detector is the simplest form of receiver that we might consider. To illustrate this point, look at Fig. 5. If you doubt the simplicity of the AM radio shown, I urge you to hook one up and give it a try. You will hear the loudest local station, so don't expect to hear the weaker ones (unless the strong one goes off the air).

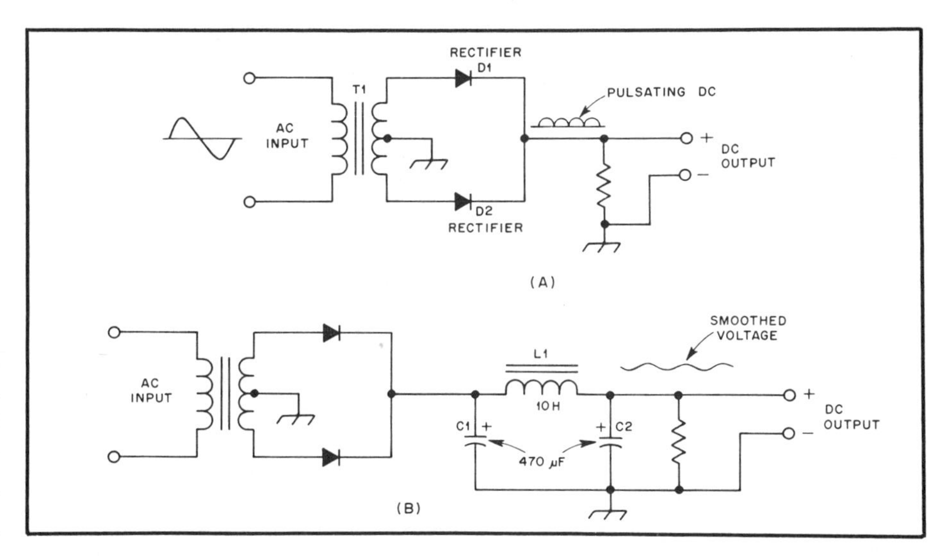

Fig. 4 — Circuit examples of diodes used as power-supply rectifiers. Pulsating dc is shown at the output of the rectifiers at A. A smoothing filter (C1, C2 and L1) has been added to the circuit (B) to minimize unwanted ripple in the dc output of the power supply.

The example of Fig. 5A has only an antenna, a diode detector (D1), a bypass capacitor and a single earphone. The more effective the earth ground and the longer the antenna wire, the louder the sound in the earphone. The circuit of Fig. 5B is a bit more complex, but will provide better performance. The combination of L1 and C1 is tuned to the desired station (you might be able to separate two or three broadcast-band stations this way). Again, the better the antenna and earth ground, the louder the signal response in the phones.

The tuned-circuit receiver is akin to the old crystal sets we mentioned earlier. The diode is coupled to L1 by means of a secondary winding that you can add. Wrap about 30 turns of fine insulated wire over the main coil winding. No. 30 or no. 32 enameled wire is suitable. You may wish to build these receivers as a workshop experiment: It is educational as well as fun.

Diode detectors are used in much more elaborate circuits than those of Fig. 5. They are used as mixers and product detectors in complex high-performance receivers, but those applications are beyond the scope of this publication. The *ARRL Handbook* contains a wealth of information on that subject; likewise with ARRL's *Solid State Design for the Radio Amateur.*

Basically, here's what happens with a detector diode. It rectifies the incoming RF signal voltage (also ac), converts it to pulsating dc voltage (as in the power-supply example) and causes the headphones to vibrate physically at an audible rate. This is caused by the pulsating dc flowing through the earphones. Simple receivers of this variety were fashioned from razor blades and a wire by U.S. soldiers during WW II. The carborundum in the razor blades would permit diode action for those "fox hole radios."

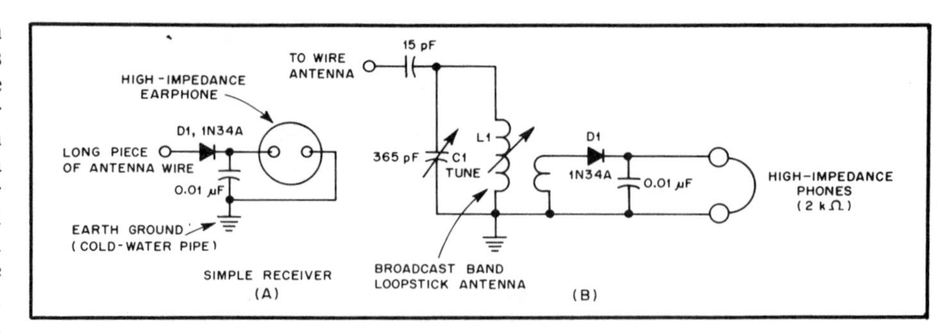

Fig. 5 — Illustration A shows the simplest form of AM receiver we might build. It consists only of a wire antenna, detector diode, capacitor, earphone and earth ground. The circuit at B has a tuned circuit that helps to separate the broadcast-band signals, but otherwise operates in the same manner as that at A.

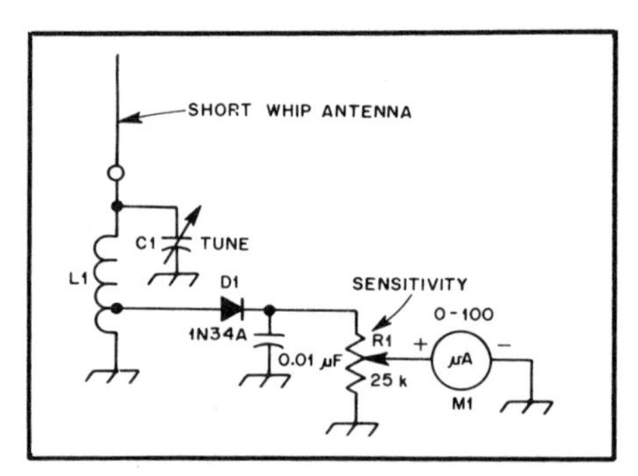

Fig. 6 — A diode detector can be used to sample RF energy. This simple field-strength meter uses rectified RF energy (pulsating dc) to cause the needle of M1 to deflect upward when C1 and L1 are tuned to the frequency of interest. R1 serves as a sensitivity control to prevent strong signals from causing damage to the meter.

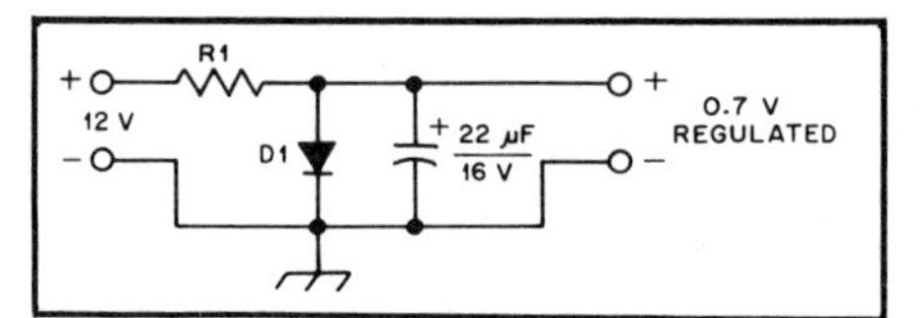

Fig. 7 — We can take advantage of the inherent barrier-voltage characteristic of a diode to provide voltage regulation (see text). R1 is chosen to limit the current through the diode junction, preventing damage to the diode.

Other Diode Uses

We can take advantage of the ac-to-dc action of the diode in creating all manner of ham-radio gadgets. For example, we might want to build a field-strength meter for tuning our transmitters or antennas. A circuit that would do the job is shown in Fig. 6. C1 and L1 are chosen to provide resonance at the frequency of interest — say, 3.7 MHz. A short whip antenna (18 to 36 inches will suffice) is attached to the top of L1 to sample the RF energy. D1 rectifies the RF energy picked up by the whip. The pulsating dc (unfiltered) flows through a sensitive meter (M1) and causes the needle to deflect upward when L1 and C1 are tuned to the signal frequency. A strong signal could harm the meter needle by causing it to become bent, so we have added a variable resistor (R1) to control the amount of dc that reaches our meter. This would be known as a sensitivity control.

Note that this circuit is not too different from the one in Fig. 5B. Instead of routing the pulsating dc into a pair of earphones, we have sent it to a meter for visual indication. D1 is connected near the grounded end of L1 (about 25% of the total number of turns) so that it will not "load" the coil and cause a broad tuning response when C1 is adjusted.

If you have an amateur license and a transmitter, why not build a field-strength meter and see how it works? You can use one of the broadcast-band ferrite loopstick antennas of Fig. 5 (Radio Shack sells them). Retain the 365-pF variable capacitor and remove turns from the loopstick until you get a meter response at 3.7 MHz when C1 is set at midrange. You don't have to tap the coil as shown in Fig. 6. Instead, add a secondary winding (like in Fig. 5B), but use only about 15 turns.

Another common use for the diode is as a regulator of dc. See Fig. 7 for a typical circuit. What's happening here? Well, we need a low-level voltage that stays relatively close to the chosen amount, despite changes in load current to the right of the diode. If we use a silicon power diode for D1 (50-V, 1-A diode, for example), it will draw considerable current and provide what is known as a minimum load current. Small changes in the circuit current along the 0.7-V output line will not cause the voltage to drop appreciably. In this circuit we are taking advantage of the diode forward-voltage characteristic, or conduction voltage. You will recall that it is 0.7 V for a silicon diode. We can add diodes in series to raise the value of the regulated voltage, adding 0.7 V per diode. Fortunately for us, special diodes are available for this general application. They are called Zener (rhymes with "keener") diodes, and they are available for regulation from a few volts to more than 100 V, and at various power levels.

More Applications

Did you know that a simple junction diode can be used in place of a mechanical tuning capacitor? Many diodes are used for that purpose in miniature equipment. In fact, modern TV receivers use tuning diodes in the receiver front end to select the channels. Fig. 8 illustrates how we might use a silicon diode for tuning purposes. When a positive voltage is connected to the cathode of a diode, it acts as a variable capacitor. This is because all diodes have a junction capacitance that changes with the applied voltage.

R2 in our circuit is used to vary the applied voltage, thereby causing changes in the junction capacitance. R1 acts to isolate the tuned circuit from ground so the performance will not be impaired. L1, C1 and D1 are the significant components of the oscillator tuned circuit. Once more we are favored with good fortune, for we can purchase special diodes designed for tuning purposes. They are available under a number of trade names, such as Varicap®

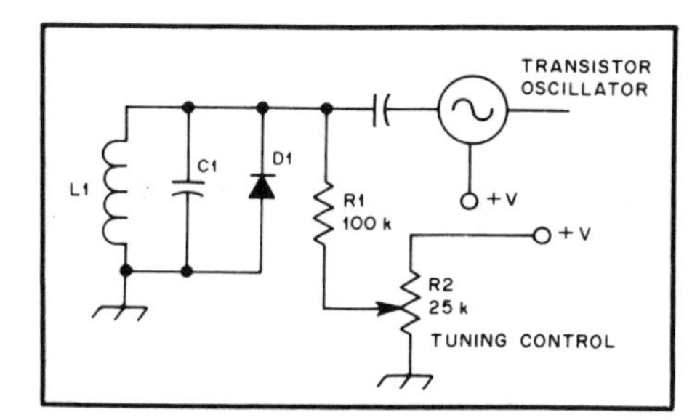

Fig. 8 — Diodes can be used to replace a mechanical variable capacitor in some instances. By applying positive voltage to the diode cathode (reverse bias), the diode internal capacitance can be made to change as the applied voltage is varied. This is an example of a tuning diode, or voltage-variable capacitor (VVC).

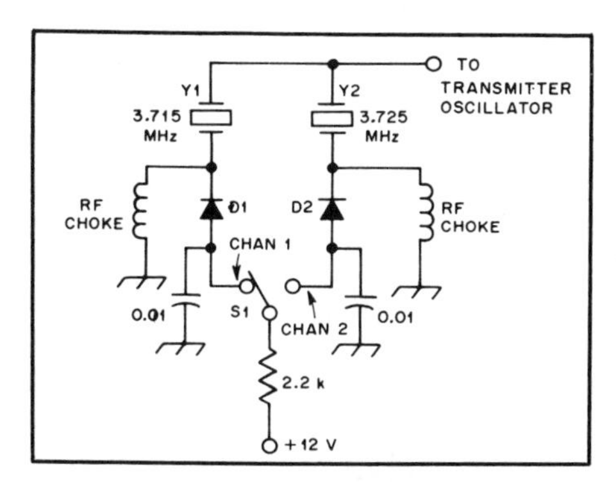

Fig. 9 — Diodes are useful also as electronic switches from dc through RF. This example shows how we might use a pair of switching diodes to select one of two or more frequency channels. When positive voltage is applied to the anode of the diode, it conducts, thereby closing the electronic switch.

and Epicap® . The generic term is *varactor diode,* which means "variable reactance" diode.

Diodes are also used as switching devices. They can be turned off and on by applying voltage and removing it, just as though we operated a mechanical switch in a voltage line. To see how this is done, look at Fig. 9. Switch leads, if long, can ruin the performance of an RF circuit. Therefore, it is convenient to use diodes as dc switches right at the point of interest in an RF circuit. The diodes can be switched by means of dc voltage that is controlled remotely with a mechanical switch (S1 of Fig. 9).

Glossary

clamp — an electrical device that limits the amount of ac voltage to a specified value.
clipper — an electrical circuit that functions like a clamp to limit the ac voltage level.
detector — a circuit that changes ac or RF energy into an audio-frequency wave that can be comprehended visually or by the human ear.
loopstick — a coil wound on a ferrite rod for use as a broadcast-band receiving antenna in small AM radios.
mixer — a device that combines two ac signals to produce a third frequency, called the intermediate frequency.
passive — a device that requires no operating voltage to make it function.
product detector — similar to a mixer except that the output frequency or intermediate frequency is at audio frequency.
squarer — an electrical circuit that converts sine waves to square waves.
Zener — a class of diodes with a constant-voltage characteristic that is useful for voltage regulators. Zener diodes are available in a variety of voltages.

When a positive voltage reaches D1 it turns on (conducts), permitting crystal Y1 to "see" a completed path to ground. Meanwhile, D2 is dormant, so Y1 is left floating, so to speak. When the diode conducts, it presents a short circuit for the current to pass through.

If we want to use Channel 2, for example, we simply move S1 to the appropriate position, thus actuating D2 and turning off D1. Diode switches are used in a great many circuits, and they can be used to switch almost any circuit we have in mind. In a similar fashion they can be placed in series with a signal line to act as "gates." When they aren't conducting, the gate is closed. When voltage is applied to them, they switch to the ON mode and the gate is opened to permit passage of the signal voltage.

Diodes can be used also as "clamps." In this type of application, they are placed between a signal line and ground to prevent the signal voltage from exceeding the barrier-voltage level of the diodes. A circuit of this kind is shown in Fig. 10. Unfortunately, the diodes cause the sine-wave audio signal to become distorted (square

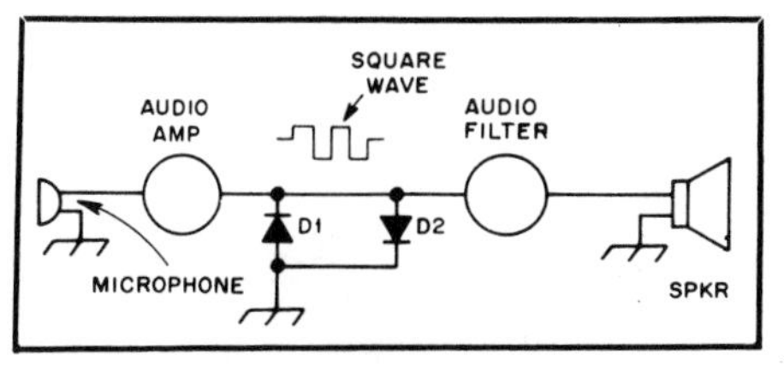

Fig. 10 — It is common practice to take advantage of the barrier voltage of a diode to limit the ac or RF voltage peaks to the barrier-voltage level. This circuit shows how we might use two reverse-connected diodes to clamp or clip the positive and negative audio-voltage peaks (to 0.7 V with silicon diodes, or 0.3 V with germanium diodes). The clipping action causes square waves (harmonics), which can cause audio distortion. A practical circuit would include a smoothing filter immediately after the diodes to restore the audio to a sine wave.

waves), which can make it sound unpleasant in the speaker. The malady can be corrected by adding an audio filter immediately after the diodes.

Clamping diodes are also called "clippers" or "squarers." When silicon diodes are used in the circuit of Fig. 10, the positive and negative peaks of audio will not exceed 0.7 V. If we desire a higher clamping level, we can place diodes in series at D1 and D2, or we can use low-voltage Zener diodes for D1 and D2.

The Wide World of Diodes

We have not given a broad picture of the diode scene in this article. There are scores of specialized diodes available for a multitude of uses. Some of the names you will hear are Schottky, IMPATT, PIN, hot-carrier, Gunn, light-emitting (LED), tunnel and solar-electric diodes. There would scarcely be enough page space in a standard-size book to describe all of the diodes and their uses. But as you advance up the technical ladder, you will recognize these special diodes and learn how to use them.

I hope you will take the time to do some additional reading about diodes. If your soldering iron is in good working order, why not pick up a few diodes and tack some circuits together? It will aid you in understanding how diodes work. Nothing beats "learning by doing."

The Magic of Transistors

Part 8: Invented by Bardeen, Shockley and Brattain at Bell Labs in 1947, the transistor has made our modern electronic world possible. Let's look at how they're used in Amateur Radio.

Doesn't everything today have transistors in it? Well, not quite. The vacuum tube remains "king of the hill" in terms of power versus cost in some applications. But, most small electronics gadgets and home-entertainment devices rely 100% on transistors or versions of the transistor (integrated circuits, or ICs).

Why is a transistor better than a tube? There are many reasons: greater reliability, increased longevity, lower cost, smaller size and reduced heating. The vacuum-tube equivalent of a 2N3904 transistor (available these days for as little as 15 cents, and smaller than a pea) would be as large as a tube of lipstick, and would cost $8 or $10 new. Furthermore, the tube would be fragile, whereas the transistor could take a pretty heavy buffeting before it became damaged. If we were to regress in the technology, and attempt to build a personal computer or a calculator from vacuum tubes, it would fill an entire living room with racks of equipment and large power supplies. I helped develop one of the first military computers in the early 1950s while employed in a research lab. Known as the MIDAC computer (Michigan digital automatic computer), it was used for BOMARC missile guidance. It filled a huge room, and stood in 6-foot equipment racks lined up side by side in 10-foot rows! The same system today could be reduced to the size of an office typewriter (without the radar display tube and electrical joysticks). Dozens of vacuum tubes were used in but one of the many circuits. Today, a single postage-stamp-size IC could be used in place of the tubes.

Fig. 1 — Left to right are low-power, medium-power and high-power transistors. There are many case styles for transistors.

What is a Transistor?

A transistor is an active semiconductor with three or more terminals. The name was derived from "transresistor" for "transfer resistor." It is made from silicon or germanium crystals that are usually formed into a junction or sandwich, as are the diodes we discussed in Part 7. The main difference is that a transistor has three elements (emitter, base and collector), whereas the diode has only a cathode and anode. The transistor can amplify signal current, but the diode cannot. Also, a transistor requires an operating voltage (it is an *active* device) for it to amplify. A diode, on the other hand, is a passive device; it does not need an operating voltage to make it rectify or clamp. It does, however, need an *applied* voltage if it is to perform a task for us. Junction transistors are commonly referred to as "bipolar transistors," sometimes abbreviated BJT (bipolar junction transistor). Fig. 1 illustrates a variety of styles and sizes of bipolar transistors.

How does a transistor compare to a tube in general terms? Look at Fig. 2 and you will observe a similarity in the symbols for the two components. Each one contains three working terminals, but the tube has two additional terminals (filaments), which are necessary for heating the cathode. Without the heaters or filaments, the tube cannot function. The transistor needs no heater. Some tubes have what are called "directly heated cathodes." They have no cathode element, and the filaments serve double duty as the heater and cathode. Those tubes reach operating conditions

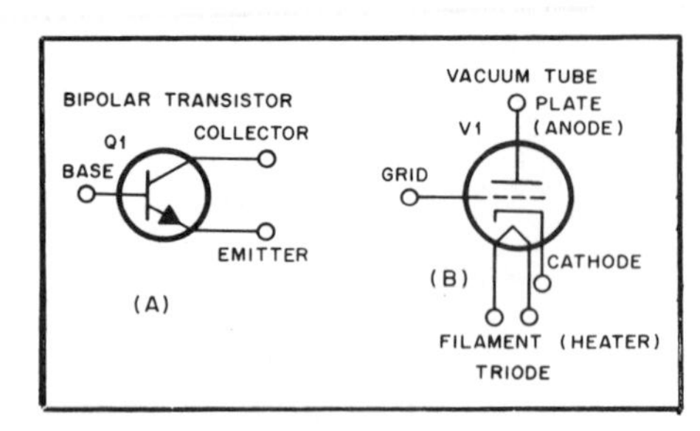

Fig. 2 — Symbol (A) for a bipolar transistor. A triode vacuum-tube symbol is included (B) for illustrating the similarity between the two triode devices.

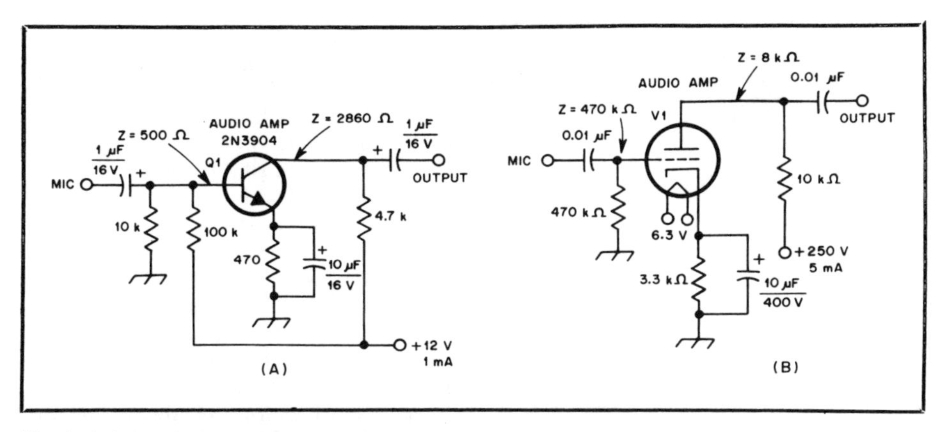

Fig. 3 — A practical circuit example for a transistorized audio amplifier (A) and a tube amplifier (B). Note the differences in the operating voltages and impedance levels (see text).

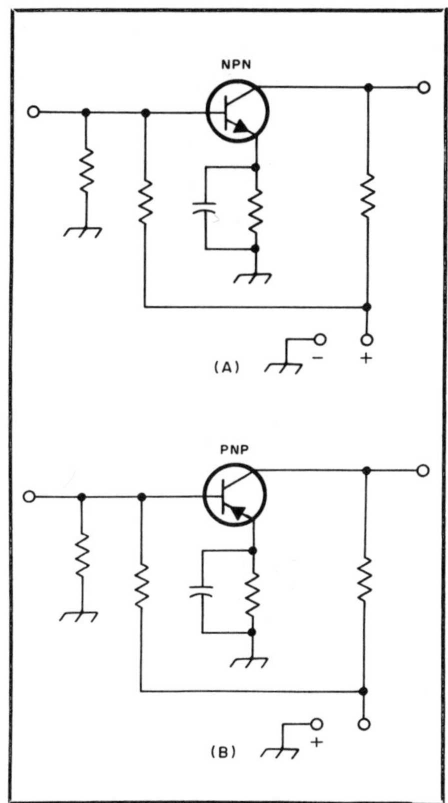

Fig. 4 — An NPN transistor (A) uses a positive collector voltage. The PNP transistor (B) requires a negative collector potential. Note the direction of the arrows for the two devices.

(from turn-on) almost as quickly as transistors do.

The example of Fig. 2B shows a triode tube that requires a warm-up time. It has separate filament and cathode elements. The transistor of Fig. 2A can be compared to the triode tube. That is, the base equates to the grid, the collector to the plate and the emitter to the cathode. Both are triodes (three electrodes), and both devices amplify ac or RF energy. The transistor amplifies current, however, while the tube amplifies voltage (ac).

Additional differences are (1) the transistor requires much lower operating voltage than does the tube, and (2) the tube has higher impedances at its terminals than does a transistor. For example, the input impedance of the transistor might present an effective impedance (ac equivalent of dc resistance) of 500 ohms at the base (base to ground), but the tube in a similar circuit could have a grid-to-ground input impedance of 1 megohm or greater. Similar comparisons can be made between the transistor collector and tube plate. Therefore, different design methods are needed for the two devices.

Let's look for a moment at Fig. 3. It shows a transistor and a tube in similar circuits. Note the differences in the operating voltages and terminal impedances. You can see there is quite a difference between the two circuits, even though they are each capable of providing approximately the same amount of amplification. The term "Z" is electronic shorthand for "impedance." You will run across this expression many times in your studies. You will observe also from Fig. 3 that the values of the resistors and capacitors are substantially different for the pair of circuits.

Additional Transistor Types

Actually, there are two types of bipolar transistor. One is called an NPN transistor, and the other is a PNP device. Symbols for the two varieties are given in Fig. 4. The NPN (negative-positive-negative) unit requires a positive operating voltage on the base and collector, but the PNP (positive-negative-positive) device needs a negative voltage on the base and collector. The distinguishing feature in the symbol that separates the two types is the direction of the emitter arrow. Observe that the arrow points out for an NPN transistor, while it points in for the PNP unit. Most transistors today are of the NPN kind, except those used for audio work. At the beginning, most transistors were PNP types, because germanium was used instead of silicon for the internal structure.

There are numerous types of tubes — some containing more than three elements. Some have four elements (two grids), and they are known as tetrodes. There are also pentodes and heptodes. In a like manner, we have transistors with an additional element. A common example is the dual-gate FET (field-effect transistor). The symbols for that and a single-gate JFET (junction FET) are shown in Fig. 5. As is the case with bipolar transistors, we have N- and P-channel FETs. The arrow in the symbols indicates the polarity of the device. At A of Fig. 5, we can see a JFET. It has an internal sandwich type of junction, as does the bipolar transistor. The dual-gate MOS (metal-oxide silicon) FET at B of Fig. 5 has a thin layer of oxide as insulating material between the gates and the remainder of the device. The drawing at C of Fig. 5 illustrates in simple terms the names of the elements within an FET. We can equate the FET to the triode tube by saying that the gate and grid are related, as are the drain and plate, and the source and cathode. The major difference between FETs and bipolar transistors is that the input impedance of the FET is similar to that of a triode tube — usually 1 megohm or greater. The effective Z is usually determined by the value of the gate-to-ground resistor used. A practical comparison between a tetrode tube and a dual-gate FET is shown in Fig. 6. We can see that the two transistor gates are used in a similar manner to the pair of grids in the tube example. A popular dual-gate MOSFET is the RCA 40673. Another is the Texas Instruments 3N211 device. When it comes to JFETs, you may recognize the

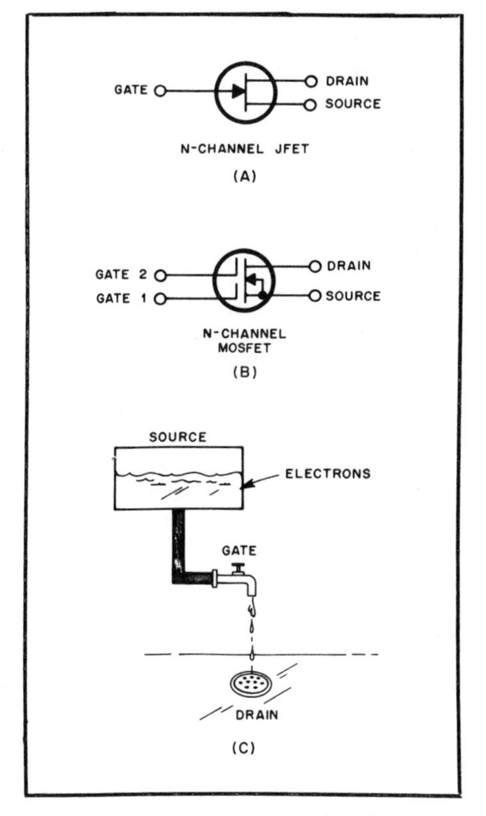

Fig. 5 — A JFET symbol is shown at A. A dual-gate MOSFET symbol is seen at B. The drawing at C shows how an FET operates.

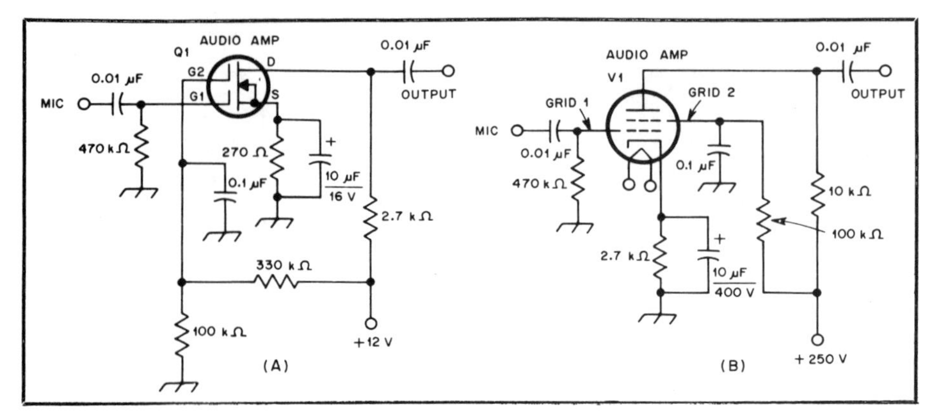

Fig. 6 — Circuit examples of a dual-gate MOSFET (A) and a tetrode tube (B) to show the similarity between the two devices.

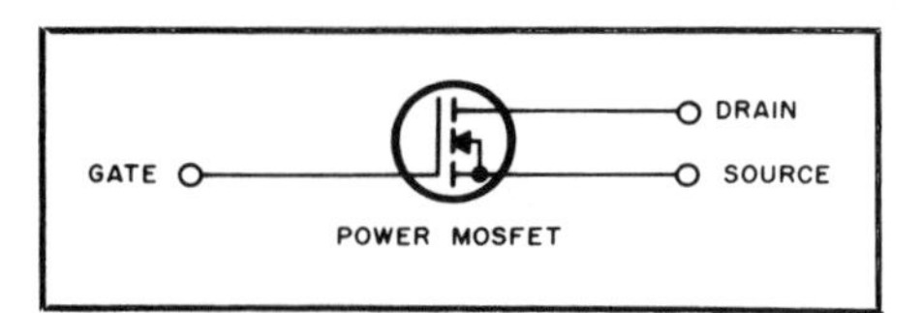

Fig. 7 — Circuit symbol for an MOS power FET of the enhancement-mode variety.

Glossary

heat sink — a metal clip or plate to which a transistor can be attached for the purpose of conducting heat away from the transistor.
heptode — a type of vacuum tube that contains seven electrodes.
JFET — a junction field-effect transistor.
MOS — abbreviation for metal-oxide silicon.
MOSFET — a field-effect transistor that uses MOS material as the gate insulation.
NPN — designator for a bipolar transistor that requires a positive base and collector operating voltage.
pentode — a type of vacuum tube that contains five electrodes.
PNP — designator for a bipolar transistor that requires a negative base and collector operating voltage.
substrate — the crystalline foundation (usually silicon) on which an IC is formed.
tetrode — a vacuum tube that has four electrodes.
thermal resistance — the effective resistance to the passage of heat between two objects bonded together.
Z — abbreviation for impedance.

number MPF102, which is an almost generic type of JFET nowadays.

Power Transistors

Thus far we have discussed only those transistors used for small-signal (low-power) applications. But, transistors can also accommodate large amounts of power. By combining many power transistors, we can build audio or RF amplifiers that deliver more than 1000 W of output power. Although no single transistor can do that job by itself, it is entirely possible to obtain more than 1000 W of output power from a single vacuum tube. It is in this area that the tube is still "king of the mountain."

There are high-power bipolar transistors and high-power MOSFETs, too. The electrical symbol for a power FET is somewhat different from that of a small FET (see Fig. 7). FETs with the three lines in place of the single drain-source line (as in Fig. 5B) are called "enhancement-mode FETs." When a single drain-source line is used it signifies a "depletion-mode FET." The difference is beyond the intent of this discussion, but it is worth mentioning to help avoid confusion.

Power transistors can generate a large amount of internal heat when they are operating. For this reason we need to use *heat sinks* to help keep them cool. Cooling fans are used on big tubes for the same reason. Excessive heat is the enemy of all electronics parts. A heat sink is a metal device that conducts the internal heat of the transistor outward. Many heat sinks are made from extruded aluminum, and they may have several rows of cooling fins on them. The transistor must be mated firmly to the heat sink to reduce "thermal resistance." Otherwise, the heat sink may be ineffective and the transistor will be destroyed. A thin layer of silicone grease is generally applied between the transistor body and the heat sink to aid the thermal bond. Some typical heat sinks are shown in the photograph of Fig. 8.

A power transistor can draw several amperes of current when a relatively low operating voltage is applied. Conversely, most power tubes require very high voltage, but draw milliamperes, rather than amperes, of current. The input and output impedances of high-power transistors are very low, often less than 1 ohm! This makes it quite difficult to work with them unless special input and output matching techniques are employed.

Combining Transistors

Everyone has heard about integrated circuits. You may think of them as large families of transistors residing under one roof. It is possible to have literally hundreds of transistors within a single IC. ICs help reduce the parts count in a circuit, leading to more-compact assemblies. The shortfall is that if one tiny internal transistor fails, the entire IC must be replaced! A number of ICs are shown in Fig. 9. ICs are available for amplifying signals to a moderate power-output level, but they are not as husky in that respect as big discrete (individual) transistors are.

ICs may contain MOSFET or bipolar transistors, or a mixture of both. They also contain diodes, resistors and capacitors. The internal workings of a simple IC are shown in Fig. 10. It is designated U1. U, the standard symbol for an IC, stands for "unrepairable." The innards we see at A of Fig. 10 are those of an RCA CA3045 transistor-array IC. Since all of the transistor leads come out of the case separately, we can use this IC in the same

Fig. 8 — Transistor heat sinks, like transistors themselves, come in a variety of shapes and sizes.

Fig. 9 — Some ICs. Each pin on the case is connected to an internal component, such as a transistor, diode, capacitor or resistor.

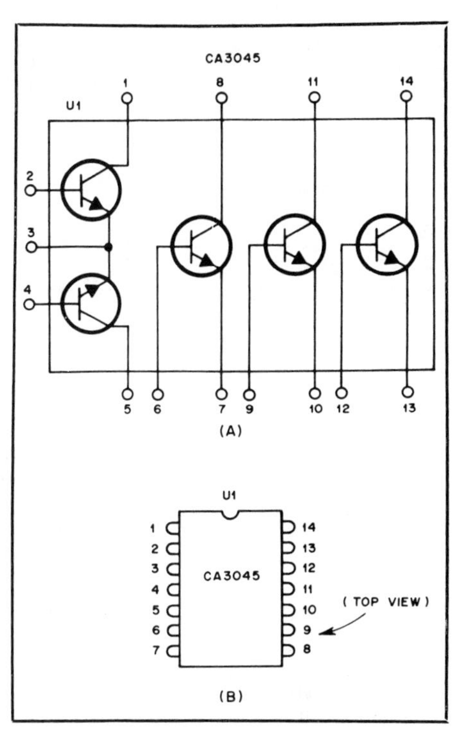

Fig. 10 — Internal circuit (A) of a simple IC. It resembles the device at B when it is enclosed in its case.

manner as five discrete transistors. Yet they are all contained in a compact assembly. The illustration of Fig. 10B is the physical format of a 14-pin, dual-in-line-package (DIP) IC. The CA3045 is one of the very simple ICs. Hundreds of transistors, resistors, capacitors and diodes can be similarly housed. The really big ICs are called LSI chips (large-scale integration). They may have as many as 40 pins coming from the case. Many LSI ICs can be found in computers and similar equipment.

There are two prominent classes of ICs. Those designed expressly for use in ac and RF circuits are referred to as linear ICs, and those meant for digital and logic applications are called logic ICs. Some hams refer to them as "analog chips" and "digital chips," respectively. The loose term "chip" refers to a piece of silicon crystal on which the IC is formed. This material is known as the "substrate."

Transistor Housing

There are numerous trappings in which a transistor may dwell. You will read about and hear mention of such things as TO-5, TO-3, TO-220, TO-92, TO-18, TO-59 and many other numbers. Don't let this confuse you. It merely signifies the physical format of the case in which the device is contained. The greater the power capability of the transistor, the larger the case it is built into. Many of the cases are designed to permit the transistor body to be mated with a heat sink. Small transistors may be in tiny metal or plastic cases, since they need no heat sinks.

Final Comments

We have skimmed the surface in our discussion of transistors. But, for those of you who are new to radio, this treatment should lay the groundwork for further learning.

Radio Antennas and How They Operate

Part 9: No amateur station can operate without an antenna. How do you pick the best type for your needs and budget? There's no simple answer, but learning their pros and cons is a good first step.

Have you strung up your "Aunt-Enna" yet? That term is sometimes used in jest to describe an antenna. But, there are less humorous names for antennas, such as *aerials* or *radiators.* The term aerial is somewhat out of style, but some old-timers still use the word on the air. You will also become aware of other antenna names as you listen to amateurs talking. For example, you might run across such words as vertical, Yagi (multielement directional antenna), rhombic (large wire antenna that is diamond shaped), and long wire.

Whatever style of antenna you select, and despite what it may be called, you will need an antenna for your ham radio station. In fact, you may eventually have several antennas on your property for communicating on various frequencies and over a variety of distances.

Most beginners start with just one antenna, but as their quest for long-distance communication (DX) increases they may add new and better antennas. Even if you don't anticipate being licensed in the near future, you will still need an antenna for shortwave listening, and for reception of W1AW code-practice sessions.[1]

Match Your Antenna

I've known a number of new hams who thought they could get on the air with a random length of wire at whatever height they could manage. Grave disappointment often followed. I have even known some Novices who gave up on Amateur Radio because "nobody answers my CQs." (The term CQ means "calling any radio amateur," and it is sent on CW or voice with the station call letters to let other hams know that we are seeking a QSO, or "contact.") These amateurs received no responses to their CQs because they had ineffective antennas, and thereby were transmitting weak signals.

The first rule for a suitable antenna is that it be *matched* to the transmitter and receiver. What does this mean? Well, our station equipment has a specific characteristic antenna-terminal impedance, generally 50 ohms. Few antennas exhibit a 50-ohm impedance without some type of adjustment or impedance-matching circuit (known also as a "matching network"). The objective is to make the antenna feed point become the same impedance as that of the station equipment. If this is not done, maximum power transfer between the transmitter and the antenna will not occur. Similarly, the received signals will be weaker than normal if the impedances are not matched. Many commercial devices are available for matching purposes. They are called Transmatches, antenna tuners or antenna couplers. In some types of systems, they can be used to match an antenna to a transmitter while in others they are used to "fool" a transmitter into thinking it is operating into a proper impedance, or *load.* This does not correct the mismatched condition at the antenna, but it does permit the transmitter to operate at full rated power without problems. From this we can establish as rule number 1 that the antenna should be matched to the feed line,

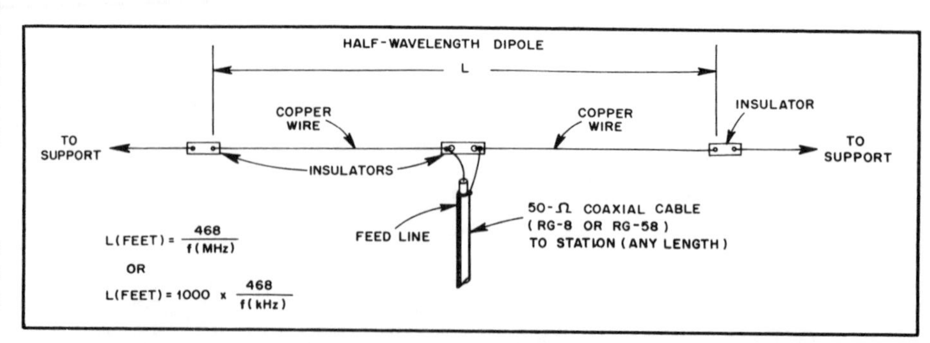

Fig. 1 — Details of a dipole antenna that is ½-wavelength long electrically. This is often a new amateur's first antenna.

[1]Check *QST* for a schedule of W1AW code-practice transmissions at various speeds.

transmitter and receiver for best results.

Some antennas need no matching device to work correctly with the station equipment. A popular example is the dipole antenna (also called a "doublet" by some hams). If cut to the proper length, it will exhibit an impedance that is close to 50 ohms. All that we need to do is connect it to the station gear through a length of 50-ohm coaxial cable. Fig. 1 illustrates this style of antenna. It is perhaps the most common "first antenna" for new amateurs. The overall length is ½ wavelength at the favored operating frequency.

For example, suppose you are a Novice and want to start operating in the 80-meter Novice band (3700 to 3750 kHz). You would want to cut the antenna for the center of that band (3725 kHz). Using the formula in Fig. 1 you would have a length (L) of 125.63 feet, or 125 feet 7½ inches. The feed line would be connected at the exact center of the wire, as shown.

A common variation of this antenna is shown in Fig. 2. What we have here is a drooping dipole, or "inverted V." The formula for length is the same as for the antenna of Fig. 1. The inverted V needs only one high support (for the center), which makes it simpler to erect. If we cut our dipoles to the operating frequency, they are said to be "resonant," and this is desirable. When an antenna is resonant, the feed point looks like a pure resistance to the feed line. If it is not resonant, we will encounter a component known as "reactance," and it can make our antenna difficult to match to the feed line. Ideally, the reactance should be tuned out or canceled by means of a matching circuit. A detailed description of reactance can be found in *The ARRL Antenna Book.*

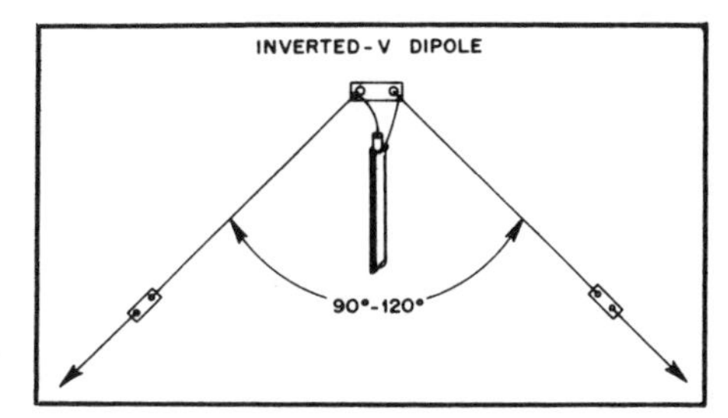

Fig. 2 — A variation of the dipole in Fig. 1. Only one supporting structure is needed for this antenna.

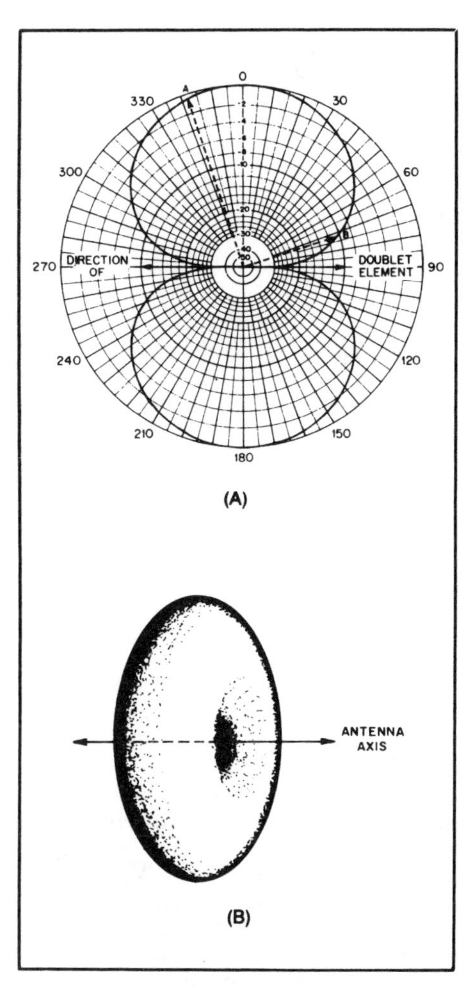

Fig. 3 — At A, the figure-8 pattern of a dipole that is ½ wavelength or greater above ground. If one could see the energy being radiated, it would have the doughnut shape at B.

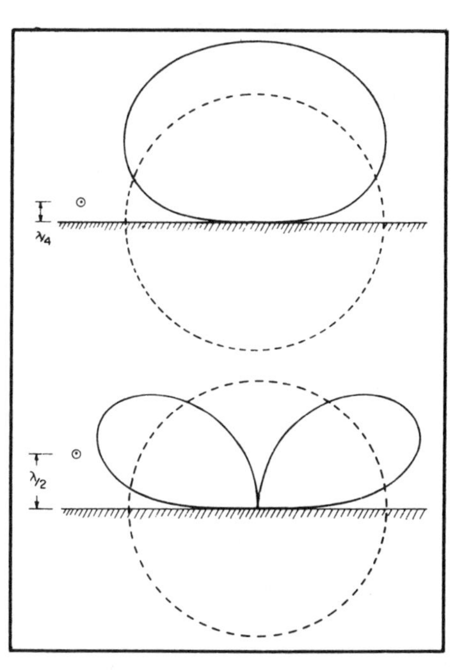

Fig. 4 — Illustrations of the radiation angles of two antenna heights (¼ and ½ wavelength). The dotted lines show what the pattern would be if there was no ground reflection under the dipoles.

What's a Dipole?

What are the characteristics of a dipole antenna? One feature is that it will radiate a figure-8 pattern (bidirectional) if it is approximately ½ wavelength or greater above ground (see Fig. 3). The lower the height above ground, the more omnidirectional it becomes. The feed impedance will vary with the height above ground, and may be as high as 100 ohms or as low as 25 ohms. In either extreme, most transmitters will work satisfactorily over these feed-impedance ranges.

It is seldom practical to erect an 80-meter dipole ½ wavelength or more above ground, for this would require support poles of approximately 130 feet! Most hams settle for whatever height is convenient, and good close-range (out to a few hundred miles) communications are common with 80-meter dipole heights of, say, 25 feet. As we mentioned earlier, low heights will cause the antenna to radiate equally well in all directions, since the figure-8 pattern tends to vanish. In general, the lower the height, the shorter the effective communications distance (depending on propagation conditions).

There is still another consideration when we deal with antenna height: a trait known as "radiation angle." It has a relationship to the angle, respective to the horizon, at which the radiation occurs. The lower the radiation angle, the greater the distance our signals can span. A dipole that is high above ground is best for low-angle work when seeking DX. On the other hand, if our dipole is quite close to ground, it will radiate a high-angle signal (almost straight up), which is good only for relatively close range, and without directional traits. We can learn from this that many factors affect dipole-antenna performance. These rules apply to all antennas erected in the horizontal format above ground.

Finally, a dipole antenna has limitations. It is good for just one amateur band when it is fed with coaxial cable. Also, it may cover only a part of an amateur band before a mismatch occurs. This phenomenon is known as "antenna bandwidth." Special broadband antennas have been designed to minimize this undesirable condition.

One way to avoid this problem is to use what is called a multiband dipole. Not only can we deal with the limited bandwidth by tuning the feed line, we can operate on many amateur bands with a single dipole erected as shown in Fig. 1 or 2. The main difference is in the type of feed line we use, plus the addition of a matching circuit in the radio room.

Fig. 5 illustrates two antennas that can be used in this manner. The antenna at A is capable of performing well from 160 through 10 meters if it is elevated well above ground and is not near conductive objects such as power or phone lines and metal structures. (All antennas work better if these conditions are observed.)

The characteristic impedance of the balanced feed line is not critical if it is in a range from 300 to 600 ohms. TV ribbon line can be used also if it is of high quality.

The antenna of Fig. 5B is known as an end-fed Zepp. It got its name from the type of antennas that were used on Zeppelins

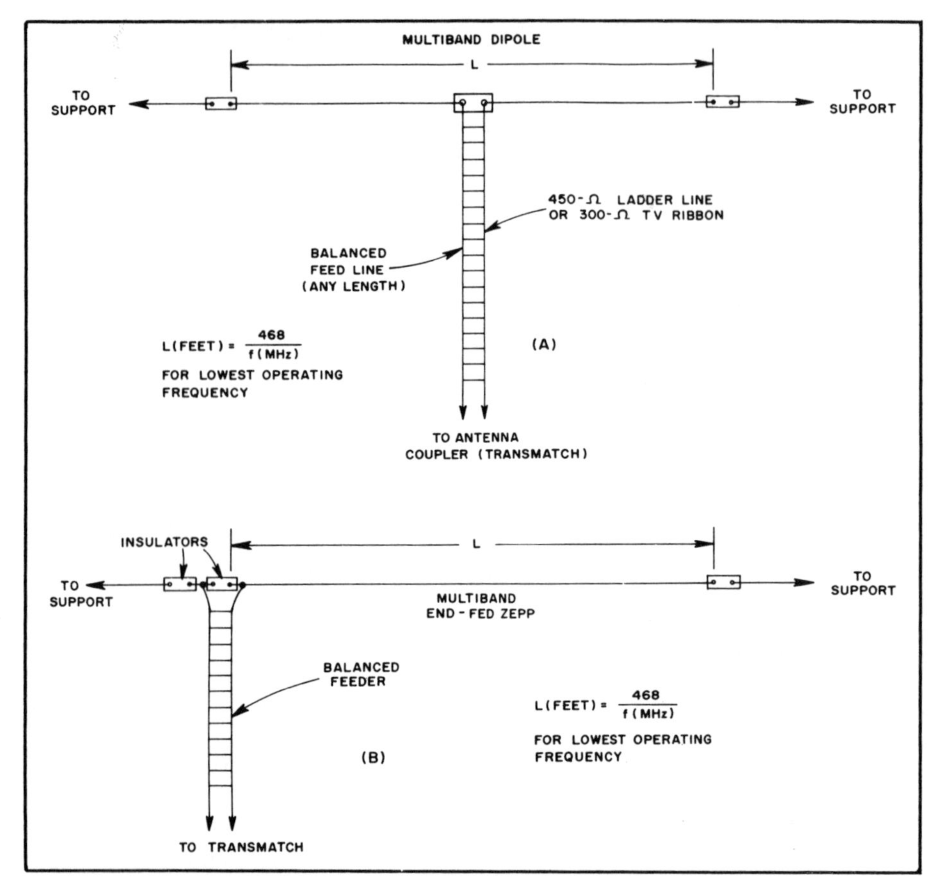

Fig. 5 — Two versions of the ½-wave wire antenna. In each example, the wire is cut to an electrical ½ wavelength in accordance with the formula in Fig. 1, using the lowest anticipated operating frequency in the calculation. A Transmatch is used to match the feed line to the station equipment.

many years ago. It is not as desirable an antenna as is the one at A of Fig. 5, but it will serve well for multiband use. The center-fed multiband antenna of Fig. 5A is probably the best choice for beginners in terms of cost versus complexity and performance.

All of the antennas we have discussed thus far are horizontally polarized. This means the radiated energy is *parallel* to the earth in the direction of the lines of force. A vertically polarized antenna has electric lines that are *perpendicular* to the earth. For close-range work (signals not reflected off the ionospheric layers), the antenna at each end of the communication circuit should be of the same polarization — vertical to vertical, or horizontal to horizontal. Horizontally polarized signals travel only very short distances over ground at HF, while vertically polarized waves can propagate long distances along the ground. Both work fine at VHF and UHF. A substantial signal loss will result if a polarization mismatch exists — vertical to horizontal antennas. For skywave (skip) communications, the polarization match is not significant because the signals that bounce off the ionosphere become somewhat "tumbled" and may arrive back at earth with various polarization traits.

Should You Go Vertical?

Antennas for use on automobiles are almost always of the vertical type. They offer low radiation angles and are simple to install. For line-of-sight VHF communications to amateur repeaters, it is necessary that the repeater antenna also be vertically polarized to prevent signal-path loss from polarization mismatch of the transmitted waves.

Many amateur fixed-location stations also have vertical antennas for HF-band use. They require very little physical space and are good low-angle radiators for DX operation. Some vertical antennas are designed for single-band use, while others contain "traps," which permit them to be used on many bands. These traps contain a coil and capacitor in parallel, thus forming a tuned (resonant) circuit for the band of interest. The trap serves to divorce electrically the portion of the antenna beyond (above) the trap. Dipole antennas can be built along the same principles to allow multiband use.

The vertical antenna radiates equally in all directions and is, therefore, an omnidirectional antenna. A ½-wavelength dipole can be erected vertically to produce the same results. Most vertical antennas are 1/4 or 5/8 wavelength long. They can be thought of as one half of a dipole. The missing half is in the earth and is called the "image half."

Fig. 6B shows how this can be envisioned. The dashed lines represent the image half of the dipole. For this to take place, we must create a ground system for use with the vertical antenna. A number of wires can be extended outward to form a circle from the base of the vertical antenna to form a ground screen. These wires may be buried in the soil, or they may lie on the ground. The more ground screen or *radial* wires used the better. But, most antenna engineers agree that 120 radials, each ¼-wavelength long, represent a reasonably good ground system for vertical antennas. However, many hams report satisfactory results when using as few as six

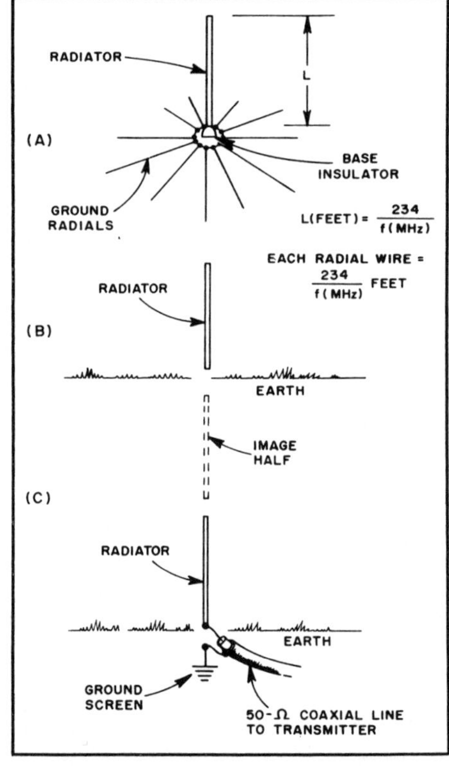

Fig. 6 — Many amateurs use vertical antennas. These radiate a vertically polarized wave, are effective over long distances, and have an omnidirectional radiation pattern (like a doughnut, with the radiator at the center). A ground screen or ground radials are needed to provide the missing half of the system, and to provide a reflective surface for the wave. Illustration B shows how the image half of the dipole appears in the earth under the radiator, making the system function as a vertical dipole. The feed line is connected as shown at C, with the shield braid of the coaxial cable connected to earth ground and the ground screen, while the center conductor of the cable attaches to the lower end of the radiator. The feed impedance is typically 30 ohms or less, so many hams insert a matching device to elevate the feed impedance to 50 ohms.

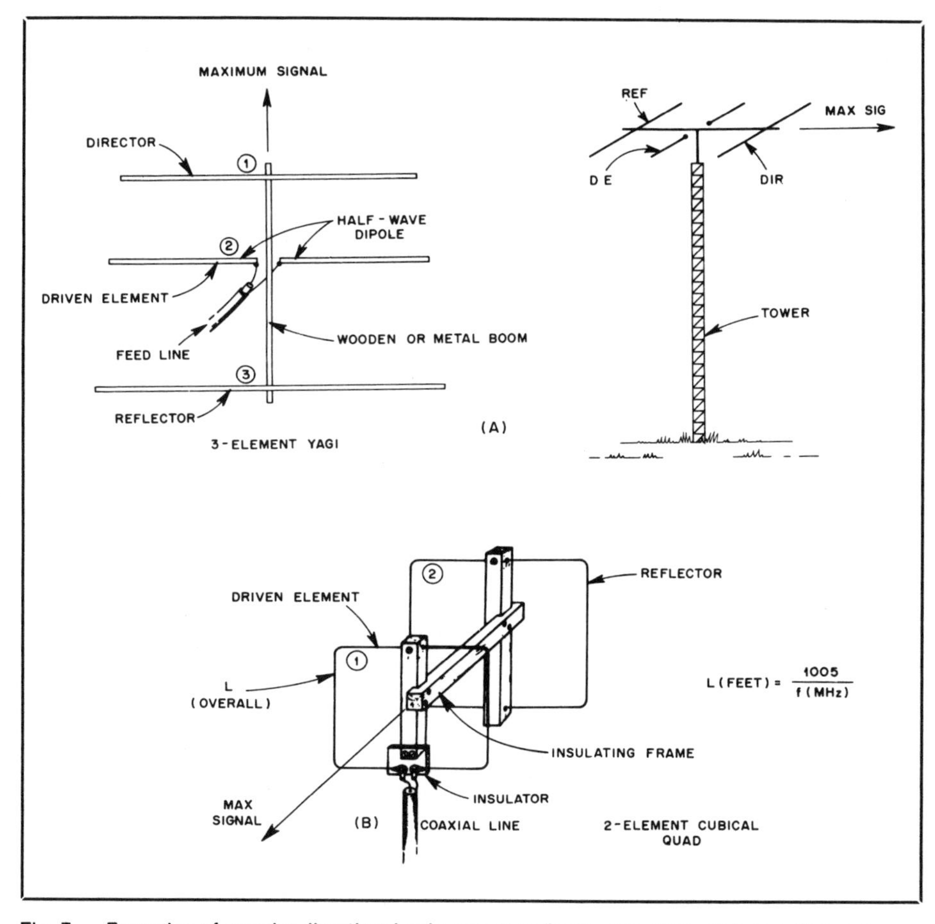

Fig. 7 — Examples of popular directional gain antennas (beam antennas). At A is a Yagi antenna with a reflector, a driven element and a director. Antenna B is a cubical quad type with two elements (reflector and driven element). Both antennas may have additional directors to increase the gain, but only one reflector is used with beam antennas.

ground radials. It depends on the ground conductivity in a given region.

Directional Gain Antennas

Many antennas have what is known as gain. Increased gain means an increase in effective signal power. If we were to replace a dipole antenna (no gain) with a 10-dB gain antenna, the effect would be the same as raising the transmitter power from 100 to 1000 W! It can be seen, therefore, that antennas with gain offer advantages for certain types of operation. Most of the popular gain antennas are also *directional,* and are erected so they can be rotated. This rotation is necessary if we are to maximize our signal power in a given direction. The directionality helps us two ways, during transmit *and* receive. Rotatable TV and FM antennas are good examples of gain types of directional antennas. These are called *beam* antennas because the signal is beamed in a particular direction, as is the case with a beam of light.

In its basic form, a beam antenna consists of a radiator (driven element) and a reflector. However, directors may also be added to increase the overall antenna gain. The two most popular forms of beam antennas are the Yagi-Uda (usually called "Yagi") and the cubical-quad antenna. Both types are shown in Fig. 7. Illustration A is a three-element Yagi antenna. The driven element (2) is a ½-wavelength dipole. The reflector (3) acts as a mirror (as with a light source) to direct the energy forward. It has a length approximately 5% greater than that of element 2. The director (1) further enhances the forward radiation of the signal. It is roughly 5% shorter than element 2. As this antenna is rotated away from the station at the other end of the communication circuit, the signal becomes weaker and weaker, and there may be several deep nulls in the response during rotation. This can work to our advantage in reducing interference from other stations and noise sources.

The cubical-quad antenna of Fig. 7B uses full-wavelength loop elements. A two-element version is shown. The reflector (2) is 5% greater in perimeter than is the driven element (1). Both antennas provide horizontal polarization, as shown. By rotating either of them 90 degrees on its axis, we can obtain vertical polarization. In other words, if antenna B of Fig. 7 had the feed terminals on the side rather than the bottom, it would have been rotated 90 degrees.

The spacing between the elements of either beam antenna will usually vary from a ¼ wavelength to somewhat less than a ¼ wavelength. The spacing chosen depends on the design objectives, which include feed impedance, forward gain and rejection off the rear of the antenna. Additional director elements can be added to these antennas to improve the gain and effective antenna bandwidth. Yagi beam antennas seldom present a 50-ohm impedance. Therefore, it is common practice to include some type of matching device at the feed point so 50- or 75-ohm coaxial cable can be used for the feed line. Antenna traps can be installed in the Yagi elements for multiband operation. Generally, the three bands concerned are 20, 15 and 10 meters. A three-band Yagi is referred to as a "tribander." Although there are numerous gain types of beam antennas, only the Yagi and the cubical quad have been treated here. *The ARRL Antenna Book* (recommended reading) describes most of the antennas used by radio amateurs, and their theory of operation. The book also explains various matching circuits.

Feeding Your Antenna

We've learned that most antennas can be used with feed lines of any convenient length. We need to be aware, however, that all feed lines have some loss, and certain ones are worse than others. And, the longer our feed line, the greater the loss. Open-wire ladder line is the least lossy, while coaxial cable of small diameter is the worst. The small coaxial lines are RG-58 and RG-59 types. Larger-diameter cable, RG-8 and RG-11, is better. The higher the operating frequency, the greater the line losses per 100 feet of cable, So it is conceivable that with a certain line loss our 100-W signal might diminish to 30 or 40 W by the time it reached the antenna — especially at VHF and higher! The same loss is experienced in the receive mode. Therefore, we should always try to keep the feed line as short as we can.

The larger conductors are best for antennas to keep losses down and to enhance the strength (physical) of the system. For wire antennas, it is wise to use no. 10 through no. 14 conductor sizes, likewise with the wire elements of cubical quads. For Yagi antennas, it is normal practice to use

Glossary

antenna — a device made of conducting material for sending transmitter signals, or for receiving them.
antenna coupler — a device consisting of coils and capacitors that can be adjusted to provide an impedance match between a transmitter and an antenna, or between a transmitter and an antenna feed line.
antenna tuner — same as an antenna coupler.
dipole — an antenna that is ½ wavelength overall and is fed at the electrical center.
director — an electrical element of a gain type of antenna, such as a Yagi.
gain — the ability of a circuit to enhance the amplitude of an ac or RF signal.
ground screen — an artificial earth ground consisting of conductive material, such as wire or screen.
image antenna — a nonphysical part of an antenna system that exists in the earth under the antenna.
ionosphere — the outer layers of the earth's atmosphere, with some electron and ion content.
load — a device that is capable of receiving power.
long wire — a straight wire antenna that is 1 wavelength or greater overall.
matched — a condition in which two identical impedances are joined.
matching network — see antenna coupler.
multiband — the property of a circuit or antenna to function on several frequency bands.
polarization — vertical or horizontal lines of force, respective to antenna radiation. Circular polarization is also possible.
radial wires — the wires in a ground screen that extend radially from the base of an antenna.
radiation angle — the angle, respective to earth, that an antenna field leaves an antenna.
radiator — the portion of an antenna system to which the power is applied; also known as the driven element in some antennas.
reflector — a conductor in a multielement gain type of antenna; it acts as an RF mirror to direct the signal forward past the driven element.
Transmatch — a matching network designed mainly for matching the transmission line (feed line) to the transmitter; similar to an antenna coupler or tuner.
trap — a resonant, parallel-tuned circuit placed in an antenna element to isolate portions of the element from other element sections.
Yagi — a type of directional gain antenna, named after the co-inventor, Hidetsugu Yagi, a Japanese engineer.

aluminum tubing for the elements. Small Yagi antennas, of the kind used at VHF and higher, may contain aluminum rods for the elements. All insulators should be of high-quality material, such as ceramic, steatite, Plexiglas®, high-dielectric plastic compounds or fiberglass.

The gauge of wire used in ground screens is not as critical. Smaller wire, such as no. 20 through 26, can be used with success, but heavy-gauge wire will stand up longer to corrosion.

What You've Learned

It may seem that I left a great deal unsaid. I did, but only because this section is meant as an introduction to the principle of antennas. Entire books are written about antennas, and still a great many details are omitted. It is for this reason that I have recommended *The ARRL Antenna Book* as a study guide.

It is important that we recognize the difference between vertical and horizontal antenna polarization, the effect of antenna height on the angle of radiation, and that the feed impedance of an antenna must be matched to that of the feed line. We also need to remember that some antennas are directional and capable of gain, while others have no gain and are omnidirectional or bidirectional.

How Receivers Work

Part 10: Antennas are of little use until we connect a receiver or transmitter to them. Understanding how receivers operate is a basic part of learning to be a radio amateur. Let's see what makes them tick.

Some hams lovingly call their receivers "hearing aids." Despite the misnomer being technically incorrect, the term does tell us what a receiver in an amateur station does: It aids us in hearing the other station's message. But, simply hearing signals does not mean we can dicipher them — at least without a good receiver (and some experience and operator skill). After all, many ham bands contain a jumble of radio signals that wax and wane, cover one another up and rattle our earphones or loudspeakers.

A good receiver is not necessarily one that costs $500 or more. Many simple, homemade receivers are capable of good performance if we are willing to do without countless knobs and features that are not essential to separating and copying signals. I think all beginners owe themselves the education and thrill of building at least one receiver. Many a new ham has been known to shout in excitement when that first distant station was pulled in on a homemade receiver. Words can't convey the feeling that goes with that experience. But, in order to pass your Novice exam or build a simple receiver, it is inportant to understand some fundamentals about receiver circuits.

There are many routes to follow in choosing a station receiver — store bought or made by hand. Let's take a look at some receiver concepts and follow briefly the evolution of the communications receiver.

At the Beginning

You may not be old enough to have heard about the "crystal set." Old-time amateurs still have nostalgic conversations about those early receivers. They consisted of a large coil on a readily available coil form, such as a toilet-tissue roll, oatmeal box or other cylindrical insulating form. The other vital element was a crystal and cat's whisker combination, which was used to detect the incoming signal. Earphones completed the package, apart from the antenna and earth ground. The circuit for such a radio is found in Fig. 1. I would like to suggest that you build one of these broadcast-band receivers for the experience. They are not suitable for reception of amateur signals, since they are incompatible with CW, SSB and FM transmissions.

Remembering that by today's standards these radios are very crude, we must accept limitations in performance. They do not separate strong signals very well, they require long antennas if one is not near a broadcast station, and the sound level in the earphones may be low on the weaker stations. But, the detected signal will be crisp and clear — more so than on some expensive receivers. The fidelity of a crystal set is amazing!

The early-day crystal radios used a galena crystal to detect the signal (Fig. 2), and adjustment of the cat's whisker was a tedious task, indeed. The experiment had to move the metal whisker about on the surface of the crystal until a "hot spot" was located. The mere act of bumping the table would require readjustment of the whisker, and the really good hot spots were seldom found a second time! The combination crystal and whisker functioned as a modern point-contact diode, which of course has no adjustment (thank goodness!). Tuning capacitors were generally not used. Instead, the insulation along one side of the main coil was bared, and a conductive slider was moved across the exposed turns to change the coil inductance, and hence alter the tuned frequency of the receiver. Other crystal sets had many coil taps that could be selected by means of a tap switch, eliminating the slider mechanism. Enough about the "dark ages." Let's learn what makes so simple a receiver operate before moving to newer things.

The signal is collected by the antenna and flows to ground through coil, L2. The combination of L2 and C1 provides resonance at the desired radio frequency (your favorite station) when you tune C1. The energy flows from the tap on L2, and as it passes through the detector diode it is rectified. This converts the radio energy (RF) from ac to pulsating dc. This dc then flows through the headphones at an audible rate, permitting us to hear the signal.

The process is not unlike that of an ac power supply that uses no filtering after the rectifier. A link winding (L1) is used over L2 when the antenna is long. Were we to

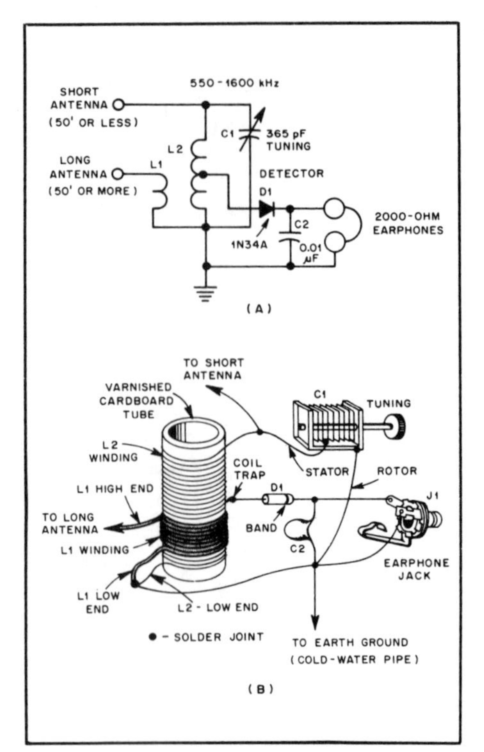

Fig. 1 — The schematic diagram at A shows the simplicity of the first radios, known as crystal sets. C1 was used to tune the stations of the standard AM broadcast band. The pictorial diagram at B illustrates how to connect the component parts of the crystal set. The radio can be built on a piece of wood or Masonite®. C1 is a 365-pF tuning capacitor. The small transistor-radio tuning capacitors are suitable for use at C1 if they have 365 to 400 pF of maximum capacitance. C2 is a disc-ceramic or tubular capacitor. D1 may be a small-signal diode, such as a 1N34A or Schottky diode of the type sold by Radio Shack. J1 is an earphone jack that matches your headphones, which should be 2 kΩ or greater in impedance. Alternatively, you may plug the output of this receiver into the input of your hi-fi amplifier. If so, insert a 0.01-µF capacitor between D1 and J1. L2 can be wound on a toilet-tissue tube. L2 should be about 220 µH in inductance. Use 150 turns of no. 26 enamel wire, close wound. Tap at 50 turns above the ground end. L1 may consist of 40 turns of no. 26 enamel wire, close wound, over the ground end of L2.

connect it directly to the top of L2, it would make the receiver tune very broadly and all of the stations would tend to come in at one setting of C1. In other words, the loading effect of the long antenna would ruin the selectivity of the receiver. The short antenna will not have a serious effect on the performance.

Enter the Vacuum Tube

The vacuum tube came along to help make reception more dramatic and practical. It permitted amplification of the radio signals *before* they were detected. This improved the receiver *sensitivity* immeasurably. Also, tubes could then be used to amplify the audio frequency (AF) after detection. This made loudspeakers practical, and several persons could listen to a radio at the same time. Radios of this class were known as tuned-radio-frequency (TRF) units. They are still used by hi-fi enthusiasts, but contain modern transistors and integrated circuits (IC).

Fig. 3 shows a typical circuit for a TRF radio. Additional RF-amplifier and audio-amplifier stages are used. They greatly increase the level of both the radio-frequency and audio signals, which enables the user to hear weak stations at comfortable volume. A shorter antenna will work reasonably well when the additional amplification is included.

Another popular amateur receiver that came along when vacuum tubes first appeared was known as the *regenerative circuit*. Exceptional sensitivity and selectivity for that era were possible with very few tubes or stages. The detector operated somewhat as an oscillator (just on the brink of oscillation). Part of the oscillator output energy was routed to the input circuit of the state and adjusted to bring the detector to the edge of self-oscillation. This action was called regeneration. It made the detector very sensitive and also aided the selectivity so that stations could be separated easily.

A typical circuit is shown in Fig. 4. The tubes (VT1 and VT2) were, depending on the era, O1As, 6C5s, 6C4s or dual triodes, such as the 6SN7 and 12AT7. Field-effect transistors, such as the MPF102, could be used today for this style of circuit. This kind of radio had a couple of problems. The detector, since it was a self-oscillating stage (like a small transistor), would permit energy to be radiated by the antenna (at the frequency to which the set was tuned). This would cause interference to nearby receivers that were tuned to the same frequency. Also, if the antenna would swing about in the wind, the radio would change frequency, causing the listener to keep his or her hand on the dial to compensate for the slight shift in frequency.

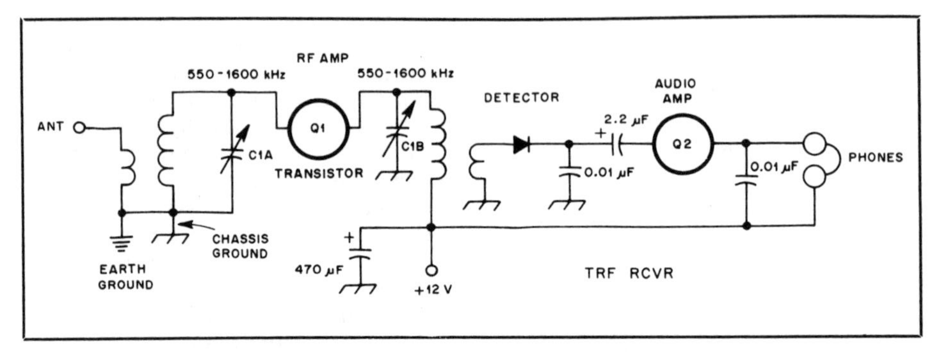

Fig. 3 — Block diagram showing how a TRF radio was set up (see text).

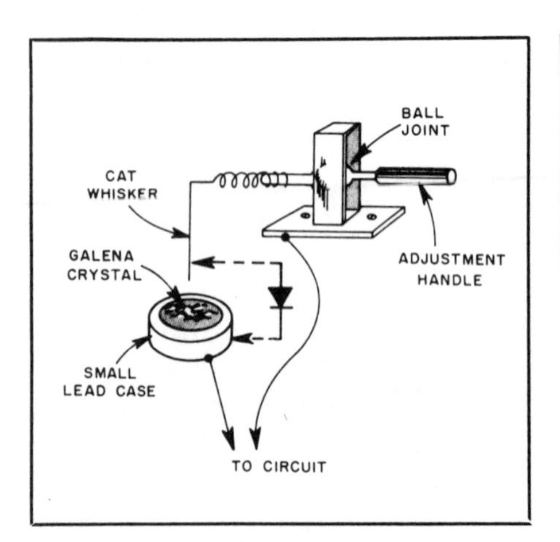

Fig. 2 — Example of a galena crystal and a cat's whisker, as used for detecting signals during the crystal-set era.

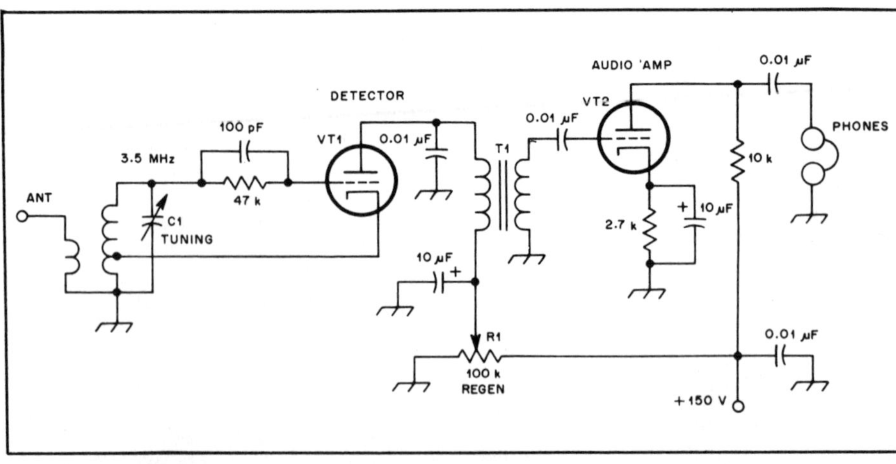

Fig. 4 — A regenerative receiver circuit. C1 was adjusted to the desired listening frequency, and R1 was set so the detector was on the verge of self-oscillation.

Clearly, something better was needed for reliable reception.

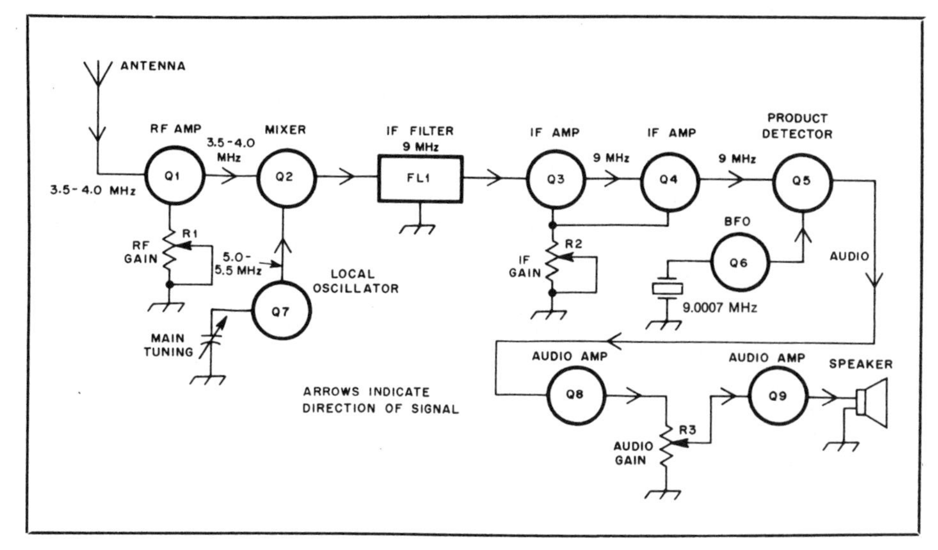

Fig. 5 — Block diagram that shows the lineup of a single-conversion superheterodyne receiver. The function of the stages is treated in the text.

The Superheterodyne Receiver

It is not my intent to saturate you with nostalgia, for as the saying goes, "What is past is past." But, the evolution of the radio is important in terms of background if we are to understand how our present-day receivers operate.

About five decades ago, we were blessed by the invention of the superheterodyne receiver concept. Strangely, it has remained the standard circuit ever since, but with improvements and frills. Many of the circuits found in early receivers are common to today's circuits. The primary advancement is the use of semiconductors in place of vacuum tubes. The solid-state parts are, in general, more efficient: They operate cooler and last longer.

What is a superheterodyne radio (often called a "superhet")? The general scheme of the critter is shown in Fig. 5. At the left, we find an RF amplifier. It builds up the signal level from the antenna and helps to separate the stations by way of *selectivity* of the tuned circuits. If the receiver includes an RF gain control (R1), it is used to vary the gain of Q1. Next, the signal, say, 3.7 MHz, is routed to the mixer, Q2. The signal from Q1 is *mixed* with the one from our local oscillator, Q7. The output from the mixer can be either the sum of the two frequencies (9 MHz) or the difference (1.6 MHz). For reasons beyond this discussion we have chosen the higher intermediate frequency (IF). The local oscillator can be thought of as a tunable low-power transmitter that creates a CW carrier. In reality, it is not a signal unless intelligence is contained on it — at least by definition. If we were to be precise in describing the local-oscillator output energy, we would call it RF voltage.

Now that we have mixed our two frequencies in Q2, we have a 9-MHz IF. To ensure that this energy is pure and free of other frequencies (including the difference IF of 1.6 MHz), we have included FL1. This filter contains four or more quartz crystals that permit the passage of the desired frequency (9 MHz) while greatly attenuating or rejecting frequencies above and below 9 MHz. Depending on the design goals for the filter, it may pass only a narrow band of CW frequencies (250 Hz), or it may be wide enough to permit SSB or AM signals to pass (2 to 3 kHz). An FM filter will pass a much wider band of frequencies (15 kHz for many modern amateur FM transceivers).

There is always some signal loss (insertion loss) through a filter, for in order for it to be a filter it must have that characteristic. The typical loss through a filter will range from 5 to 10 decibels (dB). If the station to which we are listening is running 100 W of power, a 10-dB filter loss would be equivalent to that station reducing its power to only 10 W. Therefore, we must build up the IF signal by means of IF amplifiers (Q3 and Q4).

Our ability to separate the signals has come through the selectivity of the RF amplifier stage and the filter, FL1. Therefore, the IF amplifiers do not need to have a high degree of selectivity, since the job has already been done. In fact, if we chose to use no tuned circuits between the IF amplifiers, we could design our circuit that way. Most IF tuned circuits are used to provide an impedance match between stages, rather than to increase the selectivity.

Now that we have increased the signal level from FL1, we are ready to detect or *demodulate* it. This brings us to the product detector, Q5, of Fig. 5. Generally speaking, it functions as does the mixer, Q2. The major difference is that the IF of this stage is at audio frequency rather than at RF. Therefore, the local oscillator (BFO Q6) is offset in frequency by an audio amount. For CW reception it is usually between 700 and 1000 Hz, depending on the designer's philosophy. Thus, our BFO crystal can be 700 Hz above or below the 9-MHz IF for CW reception. The offset is about 1.5 kHz for SSB reception. No BFO is needed for AM or FM reception, but special detectors are required. A product detector can be used for AM reception, however, if the AM signal is tuned in as one might tune in an SSB signal (tuned until no whistle from the AM carrier is heard). A wider IF filter is desirable for AM reception so that better fidelity will result.

Now that we have detected the signal, all that remains is to build it up (at audio frequency) until it is strong enough to operate headphones or a speaker. An audio-gain control (R3) is included for setting the level for comfortable listening. Some receivers use an IF-gain control (R2) for varying the IF gain.

The circuit of Fig. 5 is that of a single-conversion superheterodyne receiver. Double- and triple-conversion receivers are common as well. They offer some advan-

Glossary

AGC — automatic gain control. An electronic circuit that lowers the receiver gain as the incoming signal becomes stronger.

BFO — beat-frequency oscillator. It generates an RF voltage that is beat or mixed with the IF signal to produce an audible voltage or signal.

decibel — (dB) — a unit of relative power measurement.

demodulate — the process of removing signal energy from an RF or IF signal and changing it to an audio frequency.

filter — a circuit used to pass desired frequencies while rejecting unwanted frequencies.

IF — intermediate frequency, as related to superheterodyne circuits.

local oscillator — generally considered the circuit in a radio receiver or transmitter that controls the operating frequency. It is adjustable by the operator from the front panel of the equipment.

regeneration — a state that exists when the output energy from a stage (amplifier) is routed to the stage input, intentionally or otherwise. It causes the stage to self-oscillate.

selectivity — the ability of a circuit to select the desired frequency while rejecting other frequencies.

sensitivity — the ability of a receiver to extract weak signals from the internal noise of a receiver to make them discernible or readable. Based on the ratio of the inherent receiver noise to the level of a received signal.

S meter — a panel instrument on a receiver that provides visual observation of received signal levels on a relative basis.

TRF radio — a nonsuperheterodyne receiver that has tuned RF amplifiers, a detector and audio amplifiers.

tages that we won't get involved with here, but Fig. 6 shows how they differ from a single-conversion receiver.

Practically, we have two receivers in series. There are two local oscillators (Q3 and Q5), two mixers (Q2 and Q4) and two IF filters (FL1 and FL2). What we are doing is converting the signal frequency from our antenna to 9 MHz, then converting it again to a lower frequency (455 kHz). The lower frequency is known as the second IF, while the 9-MHz frequency is the first IF. A triple-conversion receiver would have three local oscillators, three mixers and perhaps another IF filter. Fig. 6 shows the most fundamental method for realizing a double-conversion receiver. Modern receivers are substantially more esoteric than the example we have examined.

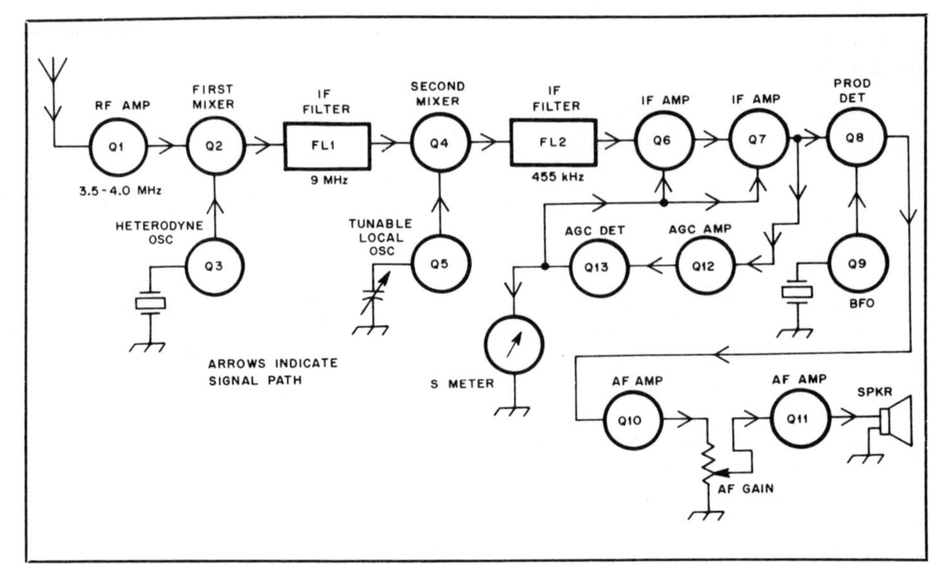

Fig. 6 — Block diagram of a double-conversion superheterodyne receiver. The function of the circuit, plus the addition of AGC and an S meter, are discussed in the text.

AGC and S Meters

Today's radios have automatic gain control (AGC) and relative signal-strength indicators (S meters). The technique for obtaining these features can be seen in simple form by returning to Fig. 6. Some IF energy is sampled at the output of the last IF amplifier, routed to an AGC amplifier (just another IF amplifier, actually), which is Q12 in our circuit, then it is rectified at Q13. The resultant dc voltage is sent back to the two IF amplifiers (Q6, Q7) for the purpose of changing their gain as the incoming signal from the antenna changes in amplitude.

The stronger the received signal, the greater the AGC voltage, and hence the lower the IF amplifier gain. This helps to keep the signal at the speaker from changing in volume, even though the received signal may vary considerably in strength. Some of the rectified AGC voltage may be used to operate an S meter, which gives us a visual indication of the relative strength of the received signal.

Today's receivers feature many additional frills, such as digital frequency readout, passband tuning, notch filters (for removing interference) fast and slow AGC response and frequency memories. But, the basic circuit is of the type shown in Figs. 5 and 6.

What Have We Learned?

If we are to summarize this lesson about receivers, we can say that the superheterodyne receiver is the common circuit today. It grew from the simple crystal detector of yesterday through a long period of evolution that brought performance landmarks step by step. A knowledge of how our receivers function is important if we are to pass the FCC license examination. It is vital also if we are to service our equipment or experience the thrill of designing and building a homemade receiver. If you wish to learn more about receivers I suggest you obtain a copy of *Understanding Amateur Radio*. There is an additional wealth of information on this subject in The *ARRL Handbook,* also available from the ARRL.

The Basics of Transmitters

Part 11: We've examined the fundamentals of receivers, so let's turn our attention toward the other half of a ham radio station — the transmitter, the amateur's on-the-air voice.

What is a transmitter? How much power must it generate to be effective? Must it be fancy in order to get the job done? Are transmitters expensive? Can I build my own transmitter? These are common questions asked by newcomers to Amateur Radio, and it is logical that the would-be ham feels a bit confused before obtaining answers to these important questions. We learned the simple ins and outs of receiver circuits in Part 10, so now we'll give similar treatment to transmitters.

The radio amateur has some options when acquiring a piece of transmitting gear. They include: (1) Purchase a new unit of commercial origin, (2) buy a used commercial transmitter, or (3) build a simple transmitter from a *QST* or *ARRL Handbook* description. The decision will be founded on how much money you can spare, whether or not you have the necessary faith in used equipment, or if you are sufficiently courageous to attempt home construction of your transmitter. I tend to favor the last choice, for as I recall my first years as a ham I recapture the thrill of talking around the world with a rig I built from scrounged and borrowed parts.

Whether you copy a design, modify one or start with your own design, there is a feeling of accomplishment that goes with the use of homemade equipment. The practicality of putting together a CW transmitter goes hand in hand with obtaining a Novice-class ham license, for CW transmitters are the least complicated and costly of the many types. Voice privileges are not available for Novices, so this makes things much simpler for the first-time builder. There are good circuits in back issues of *QST* and in the ARRL technical books.

Meet the Transmitter

In the early days of Amateur Radio, hams used what was known as a spark transmitter. By today's standards it is the most crude form of equipment for generating a Morse code radio signal. Voltage was fed to a mechanical interrupter that caused an arc when the telegraph key was closed. This wide-band energy was concentrated as much as possible in a narrow band of frequencies by means of a tuned circuit that was resonant at the desired operating frequency. The resultant note was broad and buzzy, but it could be copied. Such devices as a rotary spark gap, doorbell buzzer or Model-T Ford ignition (spark) coil were commonly used to cause the spark that became the radio signal. If we attempted to use that type of device today, our stations would interfere with every radio and TV set for blocks — or even miles! Furthermore, there would be room for only a few such signals in any of our CW bands!

After the spark transmitter was replaced by the vacuum-tube transmitter, things began to shape up in Amateur Radio. Greater distances were covered, and the ham bands could accommodate many signals at a given time. Early tube transmitters used a coil and a capacitor to control the operating frequency. This LC circuit was tuned to the desired operating frequency. Fig. 1 shows a simple version of this kind of transmitter. C1 and L1 are tuned to the operating frequency, and C1 is the main tuning control. C2 and L2 are also tuned to the operating frequency. L3 couples the output energy to the antenna system. This circuit is known as an oscillator or "LC oscillator." The key is inserted at J1. When the key is up, there

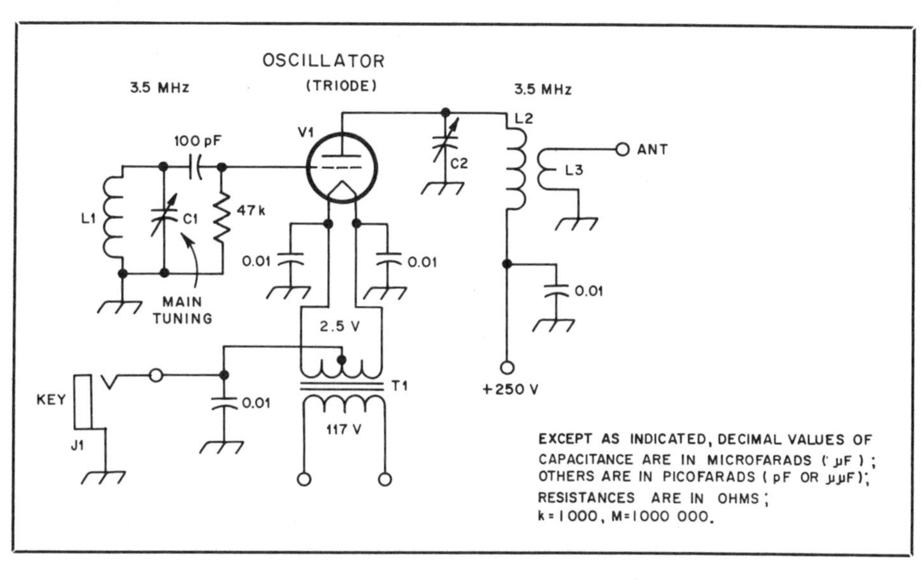

Fig. 1 — Circuit diagram of a vacuum-tube transmitter of the type used in the early days of Amateur Radio. C1 was used to change the operating frequency.

Fig. 2 — Photograph of various quartz crystals in their plug-in holders.

is no dc return to ground for the oscillator, V1, and no oscillation takes place. Key closure completes the dc circuit and causes power to be generated. Similar circuits are in use today, but not as transmitters. They may be used in some low-power part of a transmitter or receiver these days, but with semiconductors rather than tubes.

The most notable advance in transmitter technology during the early days of Amateur Radio came with the invention of the quartz crystal. It consists of a thin slab of rectangular quartz. The crystal is placed between two electrodes and enclosed in an insulating case or holder (see Fig. 2). When the crystal is excited electrically, as in an oscillator circuit, it vibrates. The operating frequency is determined by the number of times per second the quartz vibrates. For example, a 3.5-MHz crystal vibrates 3.5 million times per second. The crystal thickness determines the vibration rate. Hand grinding was the old method for crystal "tailoring," but an etching process is used today.

An example of a crystal-controlled oscillator is given in Fig. 3. It is an untuned oscillator because it has no adjustable coil and capacitor combination. It can operate only at the crystal (Y1) frequency. To change to a new frequency, we must plug in a different crystal at Y1. This circuit, like all oscillators, is basically an amplifier. But, part of the output power is routed back to the input of the amplifier to cause self-oscillation, or oscillation of the crystal.

Amplifiers should not oscillate when used strictly as amplifiers, but sometimes they do if careless design or layout permits output power to sneak back to the input side of the amplifier. This causes what is known as *instability*. So, a stage of that type becomes an oscillator, even though it is not meant to be one! The circuit in Fig. 3 is known as a Pierce oscillator — named after the man who invented it. There are many kinds of crystal oscillators, such as the Colpitts, tri-tet, Clapp, overtone and Butler. They all accomplish the same thing, but have different circuit hookups.

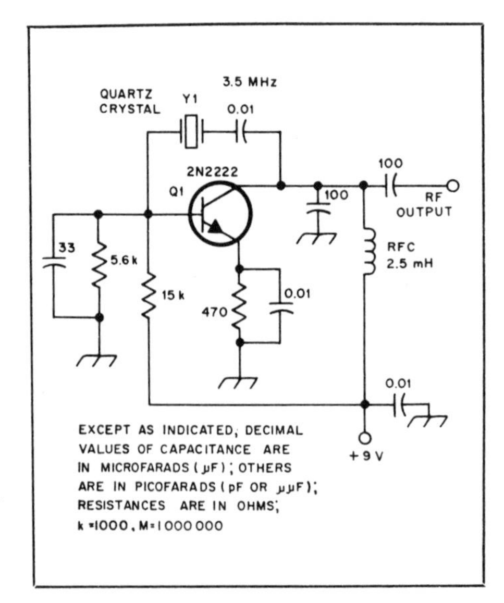

Fig. 3 — A crystal-controlled Pierce oscillator. This circuit may be used for a lab project, and can become a code-practice oscillator if used with a CW receiver.

You may wish to gather the parts for the circuit of Fig. 3 and assemble it. This will give you valuable first-hand experience concerning oscillator operation. You may hear the oscillator signal by tuning a shortwave receiver to the crystal frequency. If you open the ground connection for the 470-ohm resistor of Fig. 3 and insert a key, you may use the circuit for code practice. A CW receiver will be needed to hear the note well. Otherwise, you will hear only a thump when you key the circuit.

A Simple Transmitter

To illustrate the most simple of transmitters, let's look at Fig. 4. Here we have a one-transistor, crystal-controlled oscillator. With the parts specified in the diagram, we can expect approximately 0.25 watt (250 milliwatts) of output power. Although this may seem like too little power to communicate over anything but short distances, many hams specialize in talking around the world with QRP (low power) because it presents a challenge. This circuit, and a good antenna, can provide surprising results.

The crystal, Y1, determines the operating frequency. C1 and L1 are tuned to the operating frequency to ensure maximum power transfer to the antenna (maximize the signal output). The turns ratio on L1 and L2 is chosen to provide a proper impedance match between the collector of Q1 and the antenna feed line. Maximum power transfer can occur only when unlike impedances are matched. In other words, if the output of a transmitter has a characteristic impedance of 500 ohms and the antenna presents a 50-ohm characteristic, we would need to use some type of device (tuned circuit or transformer) to step the 500-ohm impedance down to 50 ohms.

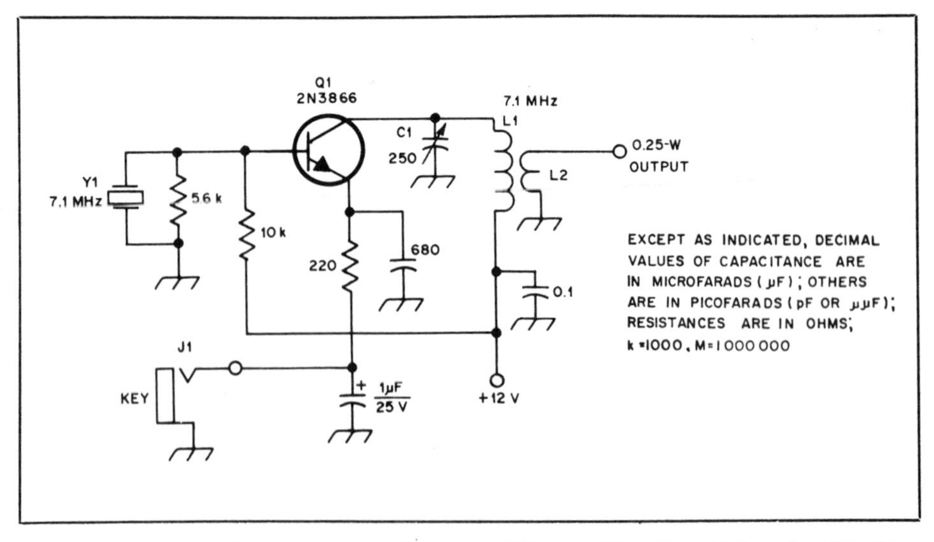

Fig. 4 — An example of a one-transistor, low-power CW transmitter. C1 and L1 are tuned to the operating frequency.

For CW operation we need only to plug our key into J1 of Fig. 4. When the key is up (open), Q1 has no dc path to ground, and it can't oscillate. When the key is down (closed), the circuit oscillates and power is delivered to the antenna. If we desire more than 0.25 watt of power, we may add one or more amplifier stages after the oscillator. The power could be increased by this means to thousands of watts if that were our objective.

Voice Operation of Transmitters

There are three common voice modes for Amateur Radio — AM (amplitude modulation), SSB (single-sideband) and FM (frequency modulation). AM was the popular mode used in the early days of radio, and remains the method used in the standard broadcast band covering 540 to 1600 kHz. The amplitude of the transmitter carrier is varied in accordance with the voice energy, and a carrier plus two sidebands (upper and lower sideband, respective to the carrier frequency) result. SSB, on the other hand, provides only one sideband (upper or lower) and the carrier is suppressed. The resulting transmitter output power varies with the voice energy, much like AM. The advantage of SSB is that the transmitter is more efficient per watt in terms of overall power consumption, the signal occupies half the bandwidth of AM and power is not wasted in generating a carrier. The narrower bandwidth reduces congestion in crowded phone bands — a matter of great importance these days with so many hams on the air.

The FM technique is somewhat different than those of AM and SSB because the voice energy is used to shift or swing the operating frequency above and below the mean carrier frequency. This shift in frequency is called *deviation*. Voice energy may be applied directly to the transmitter oscillator to create FM. Another form of FM is PM (phase modulation). The end result of either system is the same. FM receivers and transmitters are discussed in more detail in Part 17.

Representative Transmitter Arrangements

Whether a transmitter operates at VLF (very low frequency) or as a generator of microwave frequencies, the general scheme of things is the same. We must have a frequency source (local oscillator), subsequent frequency multipliers and/or amplifiers and resonant circuits. If voice operation is used, we need a *modulator*. It contains a speech amplifier and a circuit that applies the amplified audio data to the transmitter RF energy.

Fig. 5 shows a block diagram of a CW type of transmitter. We have included frequency doublers and amplifiers to provide a general idea of what might be found in a transmitter circuit. The frequency multipliers could be triplers or even quadruplers, if that would aid us in arriving at the desired transmitting frequency. On the other hand, we could design a transmitter that had no frequency multipliers: The transmitter output frequency would be the same as that of the oscillator. We might have one or two intermediate amplifiers to ensure the required excitation power to the final amplifier.

The oscillator in Fig. 5 need not be crystal controlled. Instead, we can use a VFO (variable-frequency oscillator), PLL (phase-locked loop) or a synthesizer to generate our operating or oscillator frequency. Most modern transmitters contain frequency synthesizers. They are very accurate and frequency-stable, and can be used to operate a digital frequency-readout display. In any event, all transmitters should contain a harmonic filter at the output in order to prevent the radiation of spurious frequencies that might interfere with other radio services, TV sets and FM radios.

The operating voltage for the power amplifier in Fig. 6 is processed by the modulator in order to provide amplitude modulation of the transmitter carrier. The remainder of the speech and RF stages are supplied with dc that contains no audio information. However, some transmitters use a small amount of modulated operating voltage on the stage immediately ahead of the power amplifier to ensure 100% modulation.

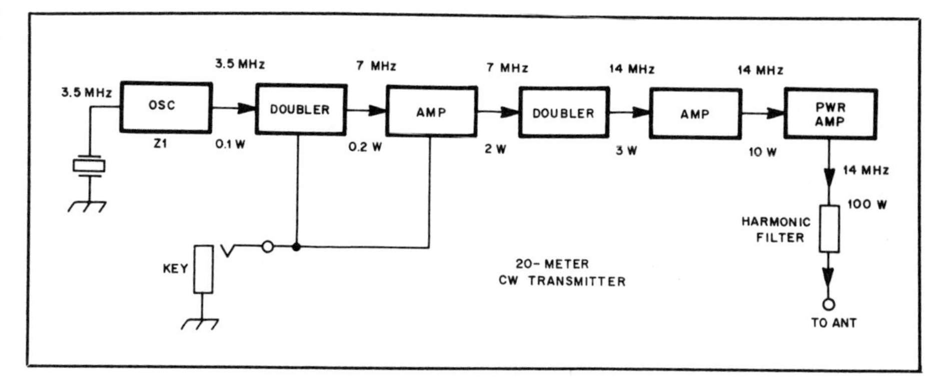

Fig. 5 — Block diagram of a simple CW transmitter with frequency doublers to increase the frequency from 3.5 to 14 MHz.

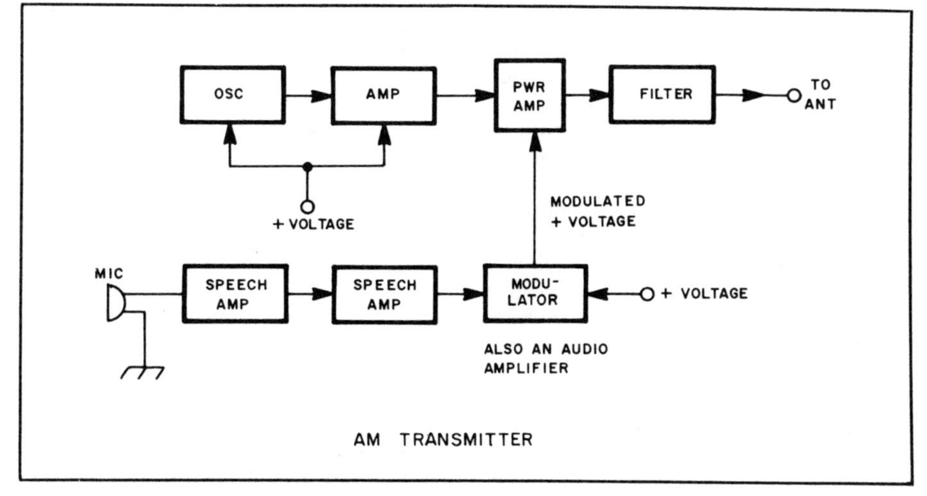

Fig. 6 — An AM transmitter is seen here in block-diagram form. The RF portion is the same as that of a CW transmitter. A modulator is used to provide AM voice output from the transmitter.

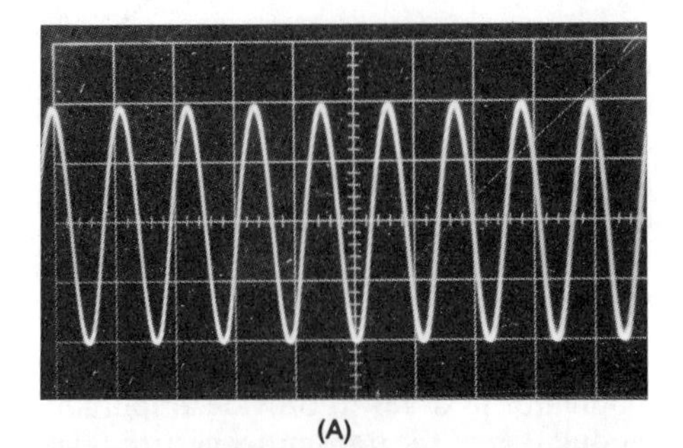

(A)

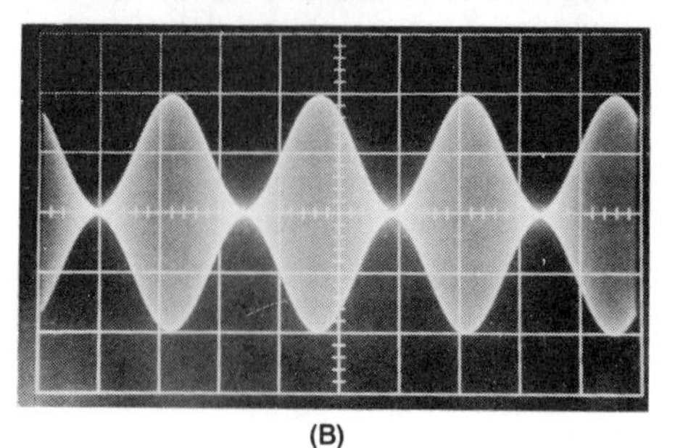

(B)

Fig. 7 — The photograph at A illustrates how an unmodulated RF wave form from an AM transmitter would appear on the face of an oscilloscope. Photograph B shows the wave form for a 100-percent-modulated carrier during AM operation.

Glossary

AM — amplitude modulation. A method of applying information to a carrier signal by superimposing the information-signal voltage on the carrier. The carrier amplitude is increased or decreased in proportion to the superimposed signal amplitude.

AMTOR — AMateur Teleprinting Over Radio. An RTTY communications method that employs error-checking procedures. Information is sent in groups of three characters (letters, numbers, punctuation, etc.). If the receiving station acknowledges correct receipt of a three-character group, the sending station transmits the next group. Should the receiving station not verify correct reception of the three-character group, the transmitting station retransmits the information. This form of RTTY overcomes the adverse effects of signal fading and interference, and ensures error-free information exchange.

ASCII — American National Standard Code for Information Interchange. A digital coding system that employs seven elements or *bits* to represent alphanumeric, punctuation and control characters. In Amateur Radio communications, the term is loosely applied to a form of RTTY communications in which ASCII is used to convey the desired information.

ATV — Amateur television. This term is usually applied to the fast-scan type of television, similar to that used by commercial TV stations. The video portion of the signal is amplitude modulated.

balanced modulator — a portion of an SSB transmitter that provides voice modulation of the carrier, then removes the carrier to leave only upper- and lower-sideband energy.

bandwidth — the amount of frequency occupied by a signal in terms of hertz or kilohertz.

carrier — the RF output from a transmitter, minus signal information.

deviation — the amount of frequency swing above and below the mean carrier frequency of an FM transmitter. The deviation is caused when audio-frequency voltage is applied to the oscillator.

discriminator — a type of detector in a receiver that has been designed for FM reception.

exciter/excitation — excitation is the RF voltage applied to an RF amplifier to cause it to amplify. An exciter is a circuit that provides excitation voltage.

FM — frequency modulation. A form of communication wherein the transmitted carrier frequency (rather than the carrier amplitude) is varied in accordance with the frequency of the modulating signal. The degree of modulation (the modulation index) is proportional to the amplitude of the modulating signal.

instability — an undesirable characteristic of an amplifier that self-oscillates.

modulator — the part of a transmitter circuit that applies modulation (audio-frequency intelligence) to one or more stages of the transmitter.

multiplier — a stage in a transmitter that doubles, triples, etc., the frequency that is applied to it.

PM — phase modulation. Another method for generating an FM signal.

ratio detector — a type of FM detector used in an FM receiver.

Repeater — A transmitter/receiver combination that receives and retransmits amateur or commercial signals at higher power levels and with a more effective antenna. This greatly increases the effective range of mobile, hand-held and/or low-power transmitters.

slope detection — a method of receiving FM signals with a standard AM receiver.

SSB — single sideband. A transmission method for creating a signal that has no carrier and just one sideband — upper or lower.

SSTV — slow-scan TV. A low-resolution, narrow-bandwidth method of picture transmission that uses audio tones applied to the modulator of a transmitter. Pictures may be transmitted in black and white or color.

synthesizer — a digital type of circuit for generating accurate, stable frequencies. Used in transmitters, receivers and test equipment. The frequencies are "synthesized" rather than generated directly by means of a crystal or LC oscillator.

Fig. 7A shows what we would see on an oscilloscope if we examined the output energy from the transmitter of Fig. 6, minus the modulation. In other words, the carrier would appear as a sine wave. But, when actuating the speech amplifier and modulator, the output wave form would appear as it is in Fig. 7B. In this example, the carrier is modulated 100% (ideal). If it is less than 100%, the signal sounds weaker in our receiver, and if the percentage is greater than 100, the signal is broad and distorted. Tubes or transistors can be used in the circuits of any of the transmitters discussed here.

Fig. 8 illustrates, in block-diagram form, the absolute basics of an SSB transmitter. The carrier is removed at the balanced modulator (balanced out, so to speak), which provides double-sideband, suppressed-carrier output to the sideband filter, FL1. Depending on the crystal used (Y1 or Y2), the output from FL1 will be upper- or lower-sideband energy, minus a carrier. The filter removes the unwanted sideband (AM transmitters transmit both sidebands, plus the carrier). The SSB energy is then routed to a mixer (as in a receiver) which is supplied in this example with 12.9-MHz energy from a local oscillator (VFO or synthesizer) to produce a sum frequency of 3.9 MHz.

Numerous other frequency schemes are popular. The one shown in Fig. 8 is but one of many combinations. The output waveform from a properly designed and operated SSB transmitter will look like that of Fig. 7B. Too high a level of modulation will cause distortion and broad signals, just as it does during AM-transmitter operation. Too little modulation will simply reduce the output power of the SSB transmitter. We should be aware that the carrier is never eliminated entirely by the balanced modulator, but it can be reduced to minus 50 dB or greater, which has the practical effect of eliminating it.

There are two methods commonly used for generating SSB signals. One is known as the filter method (Fig. 8), wherein a filter made from quartz or piezo crystals is used. In the other technique, known as the "phasing method," the unwanted sideband is removed by complex resistive and capacitive audio-phasing networks. Phasing types of transmitters have fallen out of popularity in recent years.

Other Amateur Transmission Modes

I would be remiss if I did not mention the additional transmission modes of ATV (amateur television), SSTV (slow-scan TV), RTTY (radioteletype), ASCII and AMTOR. For all practical purposes, the transmitters used for these more exotic communication modes follow the CW, AM or SSB formats described here. A proper treatment of how these modes differ from those we have already discussed would require more pages than we can devote to this article. But, you may find detailed information about these techniques, and those we have treated here, by referring to the *ARRL Handbook* and many past issues of *QST*. The ARRL technical department can provide a list of appropriate bibliographies from which to select suitable reference material. Please include an s.a.s.e. with your request.

Some Closing Thoughts

It is the intent of this article to familiarize you with the cornerstones of transmitter principles. Modern-day circuits are far more complex than the examples provided here, but the concepts are the same with regard to how the signal is generated. The schematic diagram of a typical modern

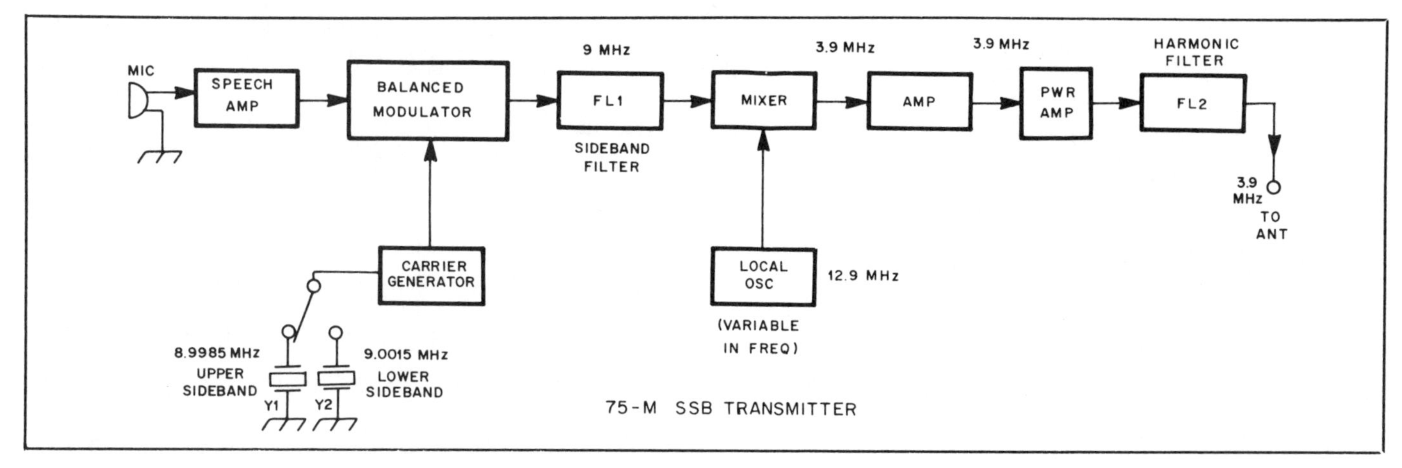

Fig. 8 — Block diagram of a single-sideband transmitter.

ham transceiver is so complicated that even *seasoned* engineers experience frustration when attempting to follow a single branch of a circuit. It would be absurd to force that kind of material on beginners, so we have followed a simplified "yellow brick road." I want to encourage you to go beyond this treatment by reading more about these principles in the *ARRL Handbook*. A few practical experiments with the oscillator circuits from this article will be beneficial, too. Good luck.

The Amateur and Electrical Safety

Part 12: Is your station as safe as it should be? If not, you may be endangering family members and neighbors as well as yourself.

Safety First! may seem a trifle boring. After all, we have heard the expression all of our lives, and we see it in print almost everywhere. In fact, "safety first" is so commonplace that we tend to become oblivious to the warning. It is a sad fact that many of us do not become aware of the dangers of high voltage and lightning until we have had personal experience with it.

We may go on our happy way with hamming for years before having a bad experience, or we may never get hooked into a voltage line that jolts us. But we must always be aware that the danger lurks constantly when we operate an amateur station. Knowledge of some specific safety measures is necessary if we are to minimize the danger of a serious accident (and pass an amateur license examination!), so let's examine the fundamentals of station safety and learn preventive measures that might save our lives.

Where Are the Hazards?

Primary among the causes of danger, or even death, in the ham shack is momentary carelessness. To help illustrate this ever-lurking specter, I will relate a personal experience that nearly cost my life. A friend, W8JEK, came to my house some years ago with a high-voltage transformer that he wanted me to test for secondary voltage. He did not have an ac voltmeter that was capable of measuring more than 1000 volts. We placed the transformer on a wooden base (for insulation purposes), connected one lead of my voltmeter to one of the secondary-winding wires, then plugged the transformer primary into the 117-V ac outlet. All seemed normal, and no smoke or strange sounds came from the big transformer. My next step could have been my last, had fate not been favoring me. I placed one hand in my pants pocket (a good safety measure) and took the remaining voltmeter test lead in my free hand. I had an alligator clip on that test lead, so decided to attach it to the remaining secondary lead of the transformer, which I did. I woke up some three minutes later on the floor of the radio room, and my mouth had the taste of acid! The last thing I remembered was feeling as though some giant had hold of my arms and was shaking me violently.

Why did this happen? The answer is lack of attention to the conditions that prevailed. First, the insulation on the meter test lead was inadequate for the amount of voltage present. Second, I was wearing shoes with leather soles (rubber is better) and was standing on a concrete floor that was damp! This is a no-no of the first magnitude! Later, we learned that the transformer secondary was rated at 2500-V, and the current capability was ½ ampere!

Needless to say, that experience was a superb teacher, and had I not been young and in good health, I'd probably not be here to write this article today. This event clearly illustrates how important it is to plan ahead — consider every possibility and ensure that every safety measure is followed *before* exposing ourselves to lethal potentials. We should always have another person present when working around dangerous ac or dc potentials: Adopt the *buddy plan* without fail!

Other common hazards are transformers that develop short circuits internally between one of the windings and the metal core and frame of the transformer. When a breakdown of this type happens, it places dangerous potentials on the equipment chassis. For this reason it is vital for us to connect a quality earth ground to all of our station gear. The ground will cause a fuse or circuit breaker to open and eliminate the safety hazard. More on this later.

Proper fusing of power supplies is similarly important to protect people from shock hazards. A fuse with too high a rating may not blow before a person is exposed to dangerous voltages.

Lightning hazards should also be considered at all times. It is unfortunate that we can do little to protect ourselves and our equipment from the tremendous voltage potential of this natural phenomenon. The best safety plan is based on preventive techniques, which we will consider later in this article.

The remaining source of danger lies in RF energy. Severe burns to the flesh can result from accidental contact with anten-

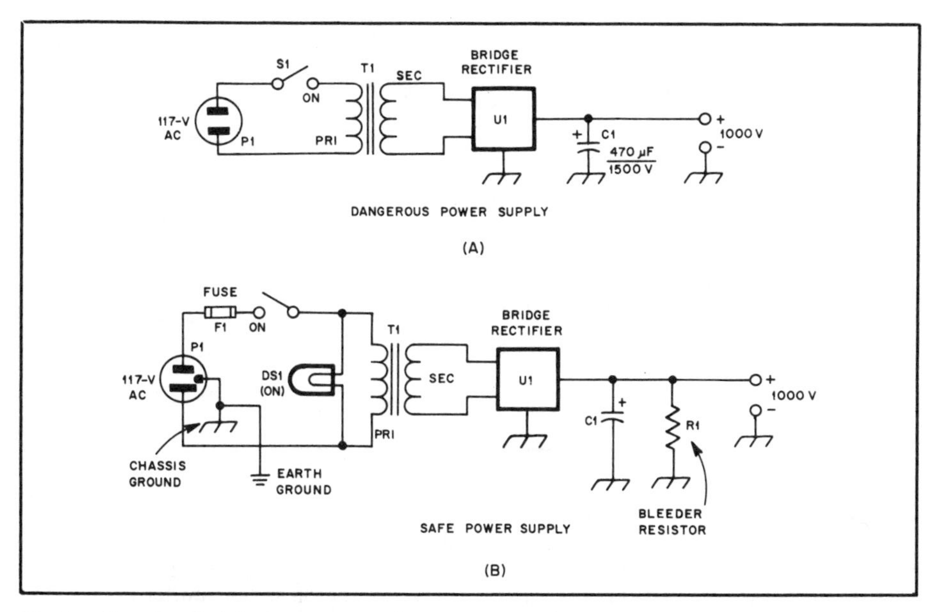

Fig. 1 — The circuit at A shows an unsafe ac power supply (see text). Example B illustrates some important safety features that should be applied to all power supplies.

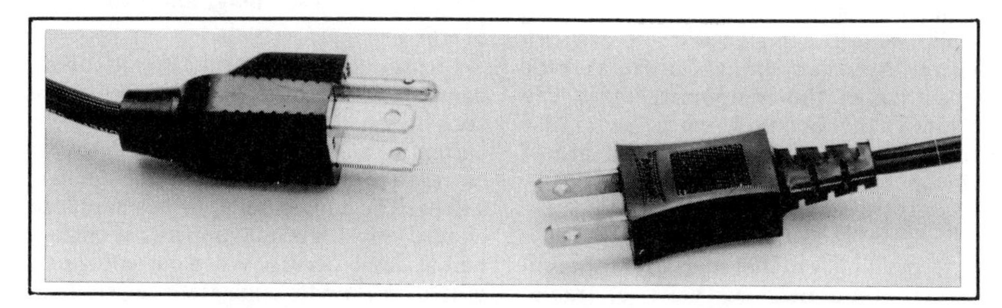

Fig. 2 — Photograph showing the difference between a three-conductor UL-approved ac line-cord plug and a two-conductor plug that is found on older equipment.

nas or transmitter components that carry high levels of RF (ac) voltage. There can even be a threat to animals with regard to RF voltage.

Power-Supply Safety

It matters not whether we use commercial ham gear or operate with homemade equipment. With the exception of mobile and certain types of portable operation, we will find ourselves relying on the ac mains for the primary power source. This dictates the need for some type of power transformer. Specific UL (Underwriters' Laboratory) safety codes should be followed. This includes a three-wire, polarized ac line cord and proper fusing of the primary side of the power supply. There is also a limitation for the distance a power supply can be from the wall outlet unless a specifically approved ac line is used.

Fig. 1 shows two simple power supplies. The first one (A) is typical of what we might find in some early-day ham shacks. Why is it dangerous? Well, first off, it does not have a safe line-cord plug (P1). The plug is not polarized (both pins are the same size and shape) and there is no third pin for automatically grounding the power-supply chassis to the ground lead in the power service.

The first circuit also lacks a fuse, which means that a breakdown in the transformer, as mentioned earlier, would permit high voltage to appear on the chassis of the equipment. Finally, there is no bleeder resistor between the output dc-voltage line and ground. A bleeder is vital for discharging or "bleeding" a power supply after it has been turned off. The filter capacitor (C1) or capacitors, depending on the design, are capable of storing a high-potential charge that could be lethal to human beings. The charge could last for hours or days, providing a significant shock hazard to persons working on the power supply or any piece of gear attached to it. The bleeder resistor drains off the stored energy within a few minutes, removing the shock probability. A bleeder resistor will, of course, dissipate some of the available power from the supply, but is a worthwhile trade-off in the interest of safety.

Fig. 1B illustrates a safe power supply. It has a three-wire power plug, a fuse, an on-off indicator lamp (DS1) and a bleeder resistor (R1). Also, as an additional safety measure, we have added a separate earth ground to the power-supply chassis. Note that P1 is polarized by virtue of one pin being larger that the others. This prevents us from plugging the line cord into the wall outlet in an improper manner. One pin goes to the neutral line and the other to the hot line. Make certain that all of your equipment contains all of these safety features. Fig. 2 shows a two-pin and a three-pin line-cord plug.

Developing a Station Ground System

A good earth ground is not a casual thing. Don't rely on a small metal rod driven into the soil. In many regions the conductivity of the soil is so poor (sand and loam) that a ground of this kind offers no effective safety measure. Furthermore, the quality of such a ground system can vary with the season, depending on the moisture content in the soil. In other words, the ground might be fairly effective during rainy seasons, but entirely ineffective in the hot, dry summer months.

How, then, might we develop a more effective earth ground? Step 1 is to connect a large-diameter conductor between the station and the nearest household cold-water line. Copper plumbing offers the best

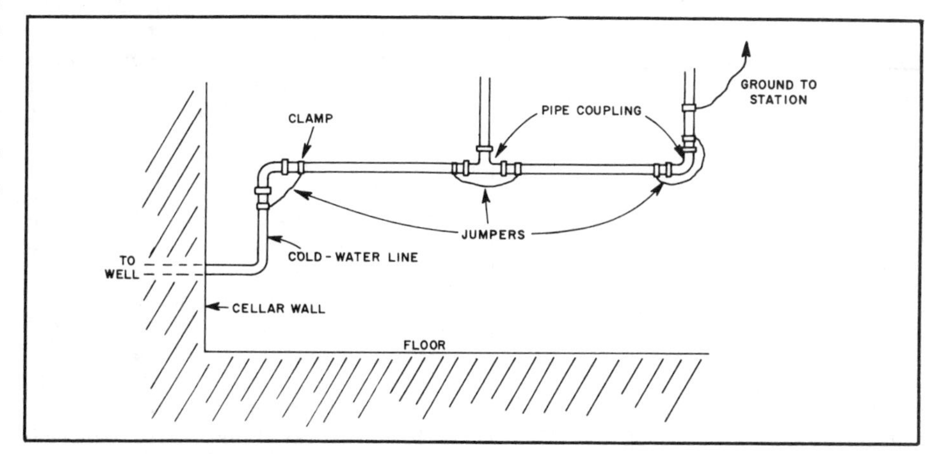

Fig. 3 — Suggested method for ensuring that good electrical continuity prevails along a length of cold-water pipe. Short conductive jumpers are bridged across each pipe joint where adaptors are present.

assurance of a quality ground, since the joints are soldered rather than being screwed together with joint compound. If iron pipes are used in your home, the problem can be solved by placing an electrical jumper wire across the pipe unions all of the way to the water source. The shield braid from RG-8 coaxial cable is good for this purpose, as is flashing copper.

The connections can be made by means of steel cable clamps around the pipes. You can use an ohmmeter to learn if the joints are resistive (bad). A good electrical joint will show a dead short when using the ohmmeter on the low-ohms range. The conductor from the water pipe to the ham shack should be a heavy conductor, such as coaxial-cable braid or similar. See Fig. 3.

Rods driven into the soil can be effective if they are installed properly. They can be used to supplement the cold-water-pipe ground. Fig. 4 contains a sketch of the method I recommend for creating an earth ground with rods or pipes. Notice that the rods (four or more) are driven into the soil to a depth of approximately 6 feet. They are arranged in a square that is 6 feet per side. Heavy conductor, such as RG-8 cable shield braid, is used to join the pipes above ground. Ideally, it should be soldered to each pipe. A propane torch is handy for this job, since a soldering iron will not develop ample heat to make a solder connection to a rod or pipe. A heavy conductor is then routed from the ground-rod cluster to the radio room. This lead should be as short as possible. Hence, the ground posts need to be placed as close to the ham shack as practicable. Galvanized pipe or copper-plated rod is suggested for the ground stakes in order to retard rusting or corrosion. Copper pipes may be used as ground rods if you can justify the cost. It may not be possible to drive a copper pipe deeply into the soil, however, since copper is relatively soft. Pilot holes could be driven beforehand with iron pipes, though.

My system has a third ground element tied into the master ground network. I have two no. 12 bare copper wires (made from stripped vinyl-covered house wire) buried 6 inches in the soil. They are 60 feet in length. One of them is attached to the base

1" PIPE OR 1/2" ROD
6'
6'
GROUND LEAD TO STATION
SOLDER ALL JOINTS
BONDING CONDUCTORS
EARTH
6'

Fig. 4 — A technique for use when an earth ground must be obtained from metal conductors driven deep into the soil. Heavy-gauge bonding straps are soldered at each rod to join the four posts electrically. The station ground is routed by the shortest means possible from the group of ground rods.

of my 50-foot tower, which is also grounded by means of rods. It is correct to say that the more extensive your ground system, the better it will be for safety reasons.

There is an additional value for a good earth-ground system: It helps minimize unwanted RF energy on the chassis of station equipment. Too high a level of stray RF energy in the ham station can cause erratic operation of the equipment, and it can "sting" the operator when he or she touches the key, microphone or cabinets of the apparatus.

The Hazards of RF Voltage

Depending on the transmitter output power, thousands of volts of radio-frequency energy can develop in the transmitter amplifier section. The antenna can also carry this high potential. RF energy may cause severe burns to the flesh if someone comes in contact with a conductor that carries it. All of our antennas should, once they are erected, be out of reach to human beings and animals.

I learned this lesson when I lived in an apartment complex where exterior antennas of any description were prohibited for aesthetic reasons. It seemed crafty for me to use the metal clothesline in my back yard for a 10-meter antenna. Each yard had one. Things worked out rather well for a month or more, until my neighbor decided on a summer evening to use the end of my wash line to support himself while he was having a lazy conversation with his wife. He chose the wrong moment, for I was working 10-meter DX at the time with a 100-watt rig! He let out a yell, which brought me to my feet. Upon investigation of the problem I learned what he had done: His hand had a burn mark across all of the palm. Fortunately for me, he understood what had caused the burn, and created no fuss. I ceased using the clothesline for an antenna!

This illustrates what can happen when an amateur antenna is close to the ground. Insulated wire may or may not prevent such a hazard. It would depend on the quality of the insulating material and its characteristic breakdown voltage rating. RF energy will, indeed, get through some inferior grades of insulation unless it is very thick. From all of this emerges a strict rule: Never work on a transmitter or antenna when the transmitter is in the operating mode.

Damage from Lightning, and Protective Measures

Lightning is the most difficult of all danger sources to deal with. Here we are considering many thousands of high-current volts. The greater the power-source current the more devastating the damage will be. The human body, for instance, can endure only a few milliamperes of current before death occurs. When current is permitted to flow through flesh, it will heat the flesh to a point of no return. This may seem like a grim statement to make in an Amateur Radio article, but it can serve as a warning that is worth remembering.

There is no complete protective means against lightning damage to personnel or station equipment! You should disconnect all antennas and ground them when they're not in use. Similarly, all ac line cords should be removed from wall outlets, since energy from lightning can enter the house via the power mains. Whenever a severe storm is forecast, cease using your ham station and follow these procedures.

Lightning arrestors can be purchased for use in amateur antenna systems, but they are by no means a fail-safe solution. I have seen a number of blown-out arrestors that were used in systems where severe equipment damage resulted. The *ARRL Handbook* shows how to build a lightning arrestor for wire antennas. It is a good idea to add one, even though it may not offer complete protection.

What Have We Learned?

All of us want to protect ourselves, our families and our neighbors against shock hazards. This suggests that we should place considerable emphasis on electrical safety when using radio gear that is powered from the ac line. Slipshod methods of grounding the station may result in getting on the air quickly, but the byproduct may be irreparable. Short, large-area ground leads attached to an effective ground system will provide the margin of safety that all of us must rely on when operating a radio station. It is worth mentioning that the better the ground system the less chance there will be for interference to nearby TV and FM receivers.

If your ham station must be located in the basement or cellar for practical reasons, use a large rubber, plastic or rubber-backed carpet pad under the area where you sit or stand near the operating desk. This will ensure additional protection against electric shock. It is best to avoid measuring high voltage until you have the proper equipment and experience. Call in an experienced fellow ham for jobs of that type.

Radio Waves and Communications Distance

Part 13: It is important to choose the correct amateur band and the right time of day for communications over great distances. Other important factors in amateur communications are the time of year and the sunspot cycle.

Most new ham radio operators are confused about which band they should use at a given time of day to communicate with certain parts of the country or the world. Have you been confused about these matters? No doubt you have pondered this subject while deciding which band to concentrate on for all-round coverage in terms of distance.

Communications over long distances, such as from the U.S. to Europe or Asia, are referred to as "DX" contacts. Since the word "distance" is relative with regard to miles or kilometers, it is best that we define DX as relating to Amateur Radio contacts over paths in excess of 1000 miles in the high-frequency bands (3.5 to 30 MHz). DX takes on a different meaning (in terms of distance) when we consider the VHF (30-300 MHz), UHF (300-3000 MHz) or microwave parts of the amateur spectrum: We may consider those frequencies above 3000 MHz as the microwave region. At VHF and higher, DX may be an appropriate term when we consider communications distances as short as 100 miles, for example. This is because the higher the operating frequency, the shorter the effective signal path over the earth's surface. This is not true of space communications, where there is acceptable attenuation (power reduction) between the transmitting and receiving antennas.

The new amateur is concerned mainly with HF (high-frequency) communications, since the Novice license is restricted to use of the 80, 40, 15 and 10-meter CW bands. A Technician class licensee has these frequencies available, along with privileges from 6 meters upward.

When you first receive your license and go on the air, chances are you will be thrilled to contact just "anybody" for the first few days. But, as you hear other hams discussing the DX they "worked" (made contact with), your appetite for DX will be stimulated! Knowledge of band characteristics for a specific time of day or year are vital if you are to be successful in talking to stations around the world. Let's examine the various parts of the HF spectrum and learn when we should use them for various communications distances.

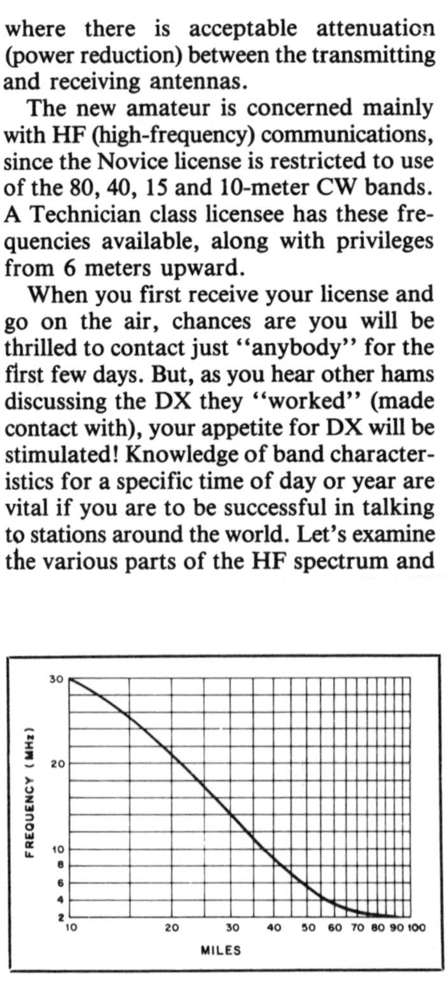

Fig. 1 — Typical high-frequency range, in miles, for ground waves compared to frequency.

Close-In Contacts

There are many occasions when we may desire to have good, solid communications across town or out to, say, 100 miles. If this is our desire, we need to select a frequency band that is best for *ground-wave* communication. A ground-wave signal is one that follows a path along the earth's surface between two antennas. The signal wave may or may not touch the ground, but it remains within the lower atmosphere during the period of travel. The lower the frequency of the HF band, the greater the ground-wave distance. Fig. 1 shows the typical ground-wave range versus frequency from 2 to 30 MHz. A vertical antenna works best for ground-wave communications. This is why commercial AM broadcast stations use vertical antennas (towers): The broadcaster wants maximum signal coverage from the station for a given transmitter output power. The amateur 160-meter band is in the MF (medium-frequency) spectrum, as are the AM broadcast stations. At 160 meters (1.8-2.0 MHz), we can expect very good ground-wave distances, compared to the bands from 80 meters through 10 meters.

Field Intensity of Waves

We have just considered the effective strength of ground waves versus frequency (Fig. 1), but we should recognize that the signal strength at the receiving antenna is measured in terms of voltage. Radio signals are very weak, so they are measured in microvolts (μV) rather than volts, as would be the case when measuring the ac from a wall outlet. A microvolt is 1/1,000,000 of a volt. Radio waves are of the ac (alternating current) type. The intensity of a signal from a transmitting antenna is always measured in terms of microvolts per meter at a distance from the antenna or signal source. The receiver S meter does not yield accurate signal-intensity readings, and is not calibrated in microvolts. An S meter is useful only for making relative measurements of signal strength, such as comparing the signal strength of two or more amateur stations, or the relative difference between two or more antennas at a given station with which you are communicating.

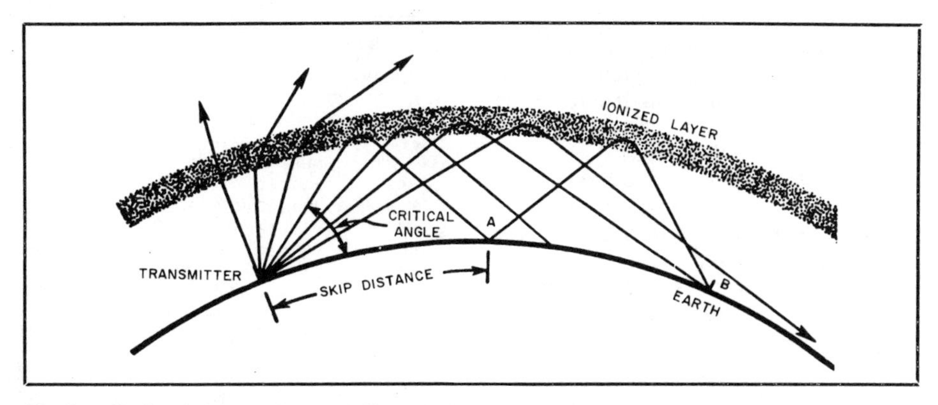

Fig. 2 — Radio signals as they are affected by the ionosphere. Some waves penetrate the ionosphere or are absorbed, while others are refracted earthward from these ionized layers (see text). Points A and B on the earth's surface in this drawing illustrate multihop skip.

Sky Waves

Distant communications may take place by means of sky waves. Sky waves travel in that area above the earth where there is no atmosphere. This region is the *ionosphere*. The condition of the ionosphere is subject to countless changes that are caused by the activity of the sun and associated changes in the earth's magnetic field. Therefore, we cannot rely on having the same sky-wave conditions from hour to hour, or from day to day. Communications by means of sky-wave propagation are often referred to as "skip communications." This is because our signals are refracted off one of the ionospheric layers and returned to earth. This is similar to bouncing a ball off a bumper in the game of pool. Fig. 2 illustrates this principle. A signal can bounce more than once, as shown.

The Ionosphere Defined

The ionosphere is a region where the air pressure is so low that free ions and electrons can move about for some time without combining to form neutral atoms. Rather lofty talk for beginners, to be certain, but I know of no other way to describe the condition. When a radio wave enters this rarified atmosphere, which is a region of numerous free electrons, it encounters a barrier, in effect, and its direction of travel is changed. This causes it to bend and deflect earthward.

Ultraviolet radiation from the sun causes the outer atmosphere to become ionized. Relatively dense areas of ionization take place, and these are called *layers*. They lie parallel to the earth's surface and occur at well-defined distances of 25 to 200 miles. Some radio waves penetrate an ionized layer deeply and then bend back toward earth. Others penetrate the layer slightly before bending downward.

Ionization is not constant within a given atmospheric layer. It tapers gradually, either side of the maximum-intensity area of the layer. The total ionization caused by the sun is never constant at a given spot for the time of day or season of the year. Because of this, there is an almost constant variation in long-distance communications effectiveness.

Ionospheric Layers

The D layer is situated 37 to 57 miles above earth. The ionization of this layer is related directly to sunlight. It commences at daybreak, peaks at noon and vanishes at sunset. During this period our 160- and 80-meter signals suffer high *absorption* loss, which limits us pretty much to ground-wave communications. At times of high solar activity (sun spots and solar flares), these bands can become completely dead. Under severe solar storms we may even find the 40-meter band severely affected. It is easy to get the false impression that our receivers are defective, for we may tune one or two bands and find no signals present!

The D layer is ineffective for refracting HF signals back to earth. Therefore, it is not useful for DX communications. We can think of it more as a nuisance than a benefit.

Now that we have properly vilified the D layer, let's look at the next layer — the E layer, some 62 to 71 miles above the earth. The E region is useful for DX work at the upper end of the HF spectrum and the lower end of the VHF spectrum. MF and lower HF signals are absorbed by the E layer in a manner similar to that of the D layer. Maximum E layer intensity occurs near the noon hour, and commences and declines in the same manner as does the D layer. The sun is not the sole ionizing agent. Ionization occurs also from solar X-rays and meteors entering the earth's atmosphere.

Our most useful layer for DX communications in the HF bands is the F layer. The height may be from 130 to 260 miles above earth. This can be compared to the E layer (see Fig. 3). The F layer may split into two layers during the daytime. If this happens, the weaker, lower layer is called F1. It is about 100 miles high and acts somewhat like the E layer. The upper F layer (F2) remains the useful one for long-distance communications. The F1 layer dissipates after sundown. DX prediction charts appear regularly in *QST*. You may also monitor W1AW bulletins for information on propagation conditions for the immediate time period.

Skip Zone and Skip Distance

Under certain critical propagation conditions there is a distance between the limits of ground wave and the beginning of skywave refracted to earth, respective to the

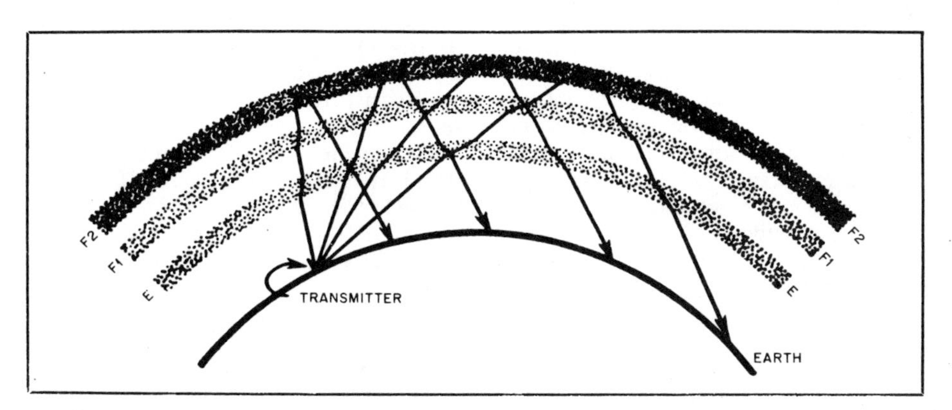

Fig. 3 — Typical daytime wave propagation at high frequency as compared to the ionospheric layers. The F2 layer is the most useful for long-range communications. The E layer is excellent for short-range skip communications at the high end of the HF spectrum and the lower part of the VHF spectrum.

location of the transmit antenna. This area between the two propagation paths is virtually dead, although there may be weak signal energy heard from the refracted wave. This ineffective communications area is called the "skip zone."

Skip distance is quite unlike the skip zone in definition. It can be described as the distance between the location of the originating signal and the point on earth where it returns to ground from the ionosphere. Therefore, with signal refraction from the F layer, the skip distance can be thousands of miles in length.

Glossary

attenuation — reduction of signal power.
band — a range of frequencies, such as 3.5 to 4.0 MHz, allocated for amateur use. The bands are designated in meters — for example, the 80-meter band.
D layer — an ionized layer in the atmosphere 37 to 57 miles above the earth.
DX — long-distance in radio communications.
E layer — an ionized layer in the atmosphere 62 to 71 miles above the earth.
F layer — an ionized layer some 130 to 260 miles above the earth.
ground wave — a signal wave from an antenna that follows the earth's surface, or slightly above the surface, for a limited number of miles.
ionosphere — the region high above the earth that has no atmosphere, but contains free ions and electrons and very low air pressure.
lobe — a concentration of radio-frequency energy that leaves a transmitting antenna and is directed toward the sky a certain number of degrees, respective to the horizon.
QRM — interference from other radio stations.
QRN — interference from atmospheric and man-made noise.
radiation angle — the angle at which a wave departs from the antenna, referenced to the horizon.
skip — the process of a radio wave bouncing off the ionosphere and returning to earth at a distant point.
skip zone — a dead signal area on earth that occurs between the limits of ground-wave signals and the beginning of the useful sky-wave signal.
sky wave — a radio wave that uses the ionosphere as a refracting medium.

Single and Multihop Propagation

We learned earlier that a refracted signal can have more than one bounce from earth to the ionosphere and back, as illustrated in Fig. 2 at points A and B. We must understand that when we send our signal into the sky it does not follow a narrow-beam-like path in the manner of a flashlight beam. Rather, it is dispersed over a wide area, and it becomes further dispersed when it is refracted from the ionoisphere. When it returns to ground it is further dispersed, becoming weaker and weaker as it hops along. For this reason, multihop propagation will usually result in weak signal readings at the distant point, even though the signal may be completely readable by the other operator.

Antenna Radiation Angle

As the radio wave is launched from our antenna, it has a particular launch angle (radiation angle), respective to the horizon. Some antennas have more than one radiation lobe (in fact, most do), and each lobe has a different intensity and radiation angle. Our concern is for the *major lobe.* The remaining lobes are referred to as *minor lobes,* but even these lobes can be used for effective communications under certain propagation conditions.

The lower the radiation angle from the antenna, the better our chances to work DX. This is because a high-angle signal may require two or more hops to reach a distant point, which will weaken the signal, as we learned while discussing dispersion. On the other hand, a very low radiation angle may enable us to work the distant station with only one hop. Launch angles between, say, 10 and 20 degrees are considered good for DX communication. The higher radiation angles are much better for shorter distances, such as we encounter at 10 and 6 meters when using the E layer for our refractive medium. Fig. 4 shows how a radiation lobe from an antenna might appear if we could see the RF energy.

The most important factor, other than the design of an antenna, is the height above ground, respective to the angle of radiation. The higher the antenna the lower the radiation angle, generally speaking. Heights in excess of 0.5 wavelength are strongly recommended. Therefore, a horizontal antenna for 40 meters should be 70 feet or greater in height. The exception is when we use a vertical antenna with a good ground system (buried or on-ground radial wires). A vertical antenna has a low angle of radiation. The trade-off is that this antenna responds equally well to signals from all directions, which can create problems from QRM (signal interference) originating in some region apart from the direction of interest. A vertical antenna is, therefore, omnidirectional in response. Furthermore, a vertical antenna is more prone to pick up man-made noise than is the horizontal antenna. If you live in an electrically noisy neighborhood, the noise (QRN) in your receiver may be so great that weak-signal reception is nearly impossible.

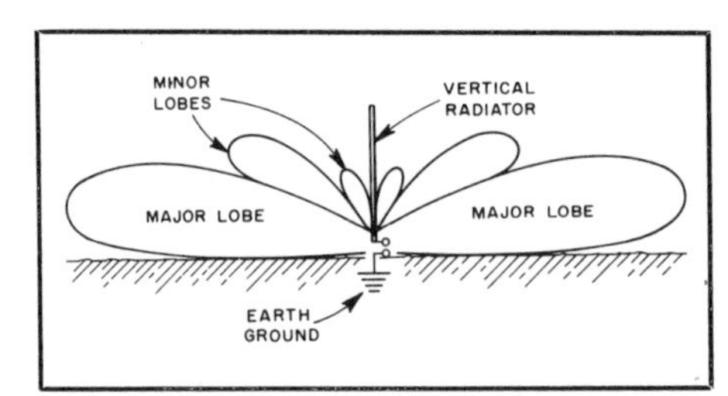

Fig. 4 — Illustration of various radiation angles versus major and minor lobes for a vertical antenna. Minor lobes occur also with most horizontal wire and beam antennas. All of these lobes are useful, depending on band conditions at a given instant and with regard to the desired communication distance (see text).

Table 1
Suggested DX Bands

Band (MHz)	Typicai Distance (Day)	Typical Distance (Night)
1.8 (160 meters)	0-50 miles	0-3000 miles
3.5 (80 meters)	0-100 miles	0-3000 miles
7.0 (40 meters)	0-1000 miles	0-3000 miles
10.1 (30 meters)	0-2000 miles	0-4000 miles
14.0 (20 meters)	0-4000 miles	0-100 miles
21.0 (15 meters)	0-4000 miles	0-100 miles
28.0 (10 meters)	0-5000 miles	0-100 miles

These distances versus time of day are based on either daylight or total darkness. Average band conditions are assumed. The actual distance worked will depend on the antenna used, the amount of transmitter power and the condition of a band at a given moment. The mileage may be greater or less than stated above. Single-hop communications are assumed here. Multihop skip will provide worldwide communications under ideal band conditions.

Horizontal antennas exhibit directivity when they are high above ground. Some have nulls off the ends (dipole antennas), while beam antennas have deep nulls off the sides and back of the array. This aids in reducing QRM from undesired directions. Man-made noise is vertically polarized, and horizontal antennas reject much of that noise since the polarization is not the same.

Sporadic E Skip

There is a form of E layer skip that is called "sporadic E." The E layer is ionized in patches rather than solidly, forming "clouds" of highly ionized atmosphere. These so-called clouds form and dissipate rapidly at times, and this is why the term "sporadic" is used. Skip from these clouds is over relatively short distances, 100 to 1000 miles. The 12-, 10-, and 6-meter bands are affected the most by sporadic E skip. However, the useful effects of sporadic E have extended as high as 148 MHz at times. I experienced this while living in Connecticut some years ago: I worked a WØ station in Minnesota on 2-meter SSB while running 10 watts to my 10-meter Yagi antenna!

Best Bands Versus Time of Day

It is not possible to produce a list of bands, effective communication distances and ideal times of day. We can only generalize because of the continual change in solar activity. We can, however, suggest the bands on which to concentrate for working local or DX contacts, day versus night. This data is presented in Table 1 for your assistance in setting up your antennas for favored bands.

Summary

Your best source of detailed information concerning the ionosphere and radio propagation is *The ARRL Antenna Book.* I have attempted here to provide a simplified, plain-language introduction to the propagation phenomenon, and to suggest steps you can take to make your first on-the-air experience a pleasant and rewarding one. Also, you will need some knowledge in this subject area if you are to pass your amateur license examination. The mysteries of the sky are many, and we have ignored a host of them in the interest of keeping this article short. I encourage you to engage in further study of this fascinating subject.

Understanding TV and Radio Interference

Part 14: Ham radio interference to home entertainment devices is a matter we can't dismiss easily. Fortunately, there are simple steps we can take to solve most problems caused by our station equipment.

In Part 13 we examined radio-wave propagation with respect to the ionospheric layers. We did not cover the effects of radio-frequency energy in the immediate vicinity of our amateur stations — the region where intense levels of RF energy are generally present when our transmitters are operating. It is not uncommon for these strong fields to create interference in nearby TV, AM and FM receivers. This near-field RF energy can also affect the performance of telephones, computers and other electronic devices found in homes.

Our responsibility as hams is to ensure that our radio equipment is not the fundamental cause of RFI (radio-frequency interference) or TVI (television interference). Often, a large part of the interference problem is the fault of the home-entertainment device, rather than the amateur's transmitter. Unfortunately, the neighbor who experiences an interference problem is hesitant to believe his or her apparatus is deficient. Often the complainant will say, "It has to be you! After all, you have that big antenna in your yard!" Such a person might also say, "It can't be my hi-fi system, I paid $1500 for it."

When the home-entertainment device is responsible for the interference problem, we need to put on a diplomat's hat and assume a new role. Animosity solves no problems, so we must try to cooperate with the irate neighbor in solving the dilemma. Let's look at the basic causes of interference, and learn what the usual steps are toward solving the problems.

Keeping Our Stations "Clean"

The first responsibility of an amateur operator is to make certain that the transmitter does not radiate harmonic energy. A harmonic is a frequency multiple (odd or even) of the operating frequency. For example, the second harmonic of 3725 kHz is 7450 kHz, the 11th harmonic is 40.975 MHz. The higher-order harmonics fall into the FM and TV bands. If they are strong enough, they can wipe out the TV picture and sound, or blot out an FM station. If these interfering harmonics are radiated by the transmitter directly or via the antenna system, they may be strong enough to cause interference a block or more away!

All transmitters generate harmonic energy. The FCC requires that all commercially made amateur transmitters for the HF bands have all spurious output energy suppressed 40 decibels (dB) or more below the peak power-output level without exceeding the power level of 50 mW. Therefore, if our transmitter puts out 100 W at the desired frequency, all harmonics and other spurious energy must be 10 milliwatts (mW) or less. At VHF, the spurious energy from the transmitter must be at least 60 dB below the peak output power. Proper transmitter design plus suitable harmonic filters can make this possible. Many homemade transmitters do not meet these performance standards, owing to incorrect design procedures and/or a lack of harmonic filtering. The ARRL, however, requires that all published transmitter circuits comply with the FCC regulations before they can appear in *QST*. Similarly, most manufactured transmitting equipment is tested for compliance before it can be advertised in *QST*. These tests are performed at ARRL Hq. in the Technical Department laboratory.

If you have an offensive transmitter, you can add an external harmonic filter to the transmitter. We'll discuss this, and other clean-up measures, later in the article.

Other Interference Causes

It is possible to have a clean transmitter, but your station may still be the cause of RFI or TVI. How can this happen? Let's suppose that somewhere in your antenna or feed line there is a poor electrical joint. In fact, a loose coaxial-cable connector may even be the culprit. A poor solder or mechanical joint can act as a rectifier diode.

When this happens, the diode-like joint generates strong harmonic currents, and these can be radiated by the antenna. A poor antenna connection may simply "sputter" and arc while you are transmitting, and this will raise havoc throughout the neighborhood.

A poorly conducting joint need not be in the antenna system. It might occur between sections of rain gutter, metal fencing or some other nearby conductor. If a sufficient amount of your RF energy is induced into such objects, they can generate and radiate harmonics.

There is a reverse-interference problem that can result from an unwanted rectifying joint: You may hear all manner of unwanted broadcast band or other commercial signals popping up here and there in the tuning range of your amateur receiver! The poor outdoor joint may be rectifying energy from an AM station. The harmonics caused by this action may fall in the amateur bands.

As an example, suppose there was an AM station near you, and the operating frequency was 1240 kHz. The third harmonic would fall at 3720 kHz — right in the 80-meter Novice band! Of course, you might hear the third harmonic anyway, if the antenna tower of the AM station was very close to your location. This would not mean that the broadcast station had a faulty transmitter. Let's consider a typical 50,000-W AM station. By law, the harmonic signal amplitude must be 40 dB or more below peak fundamental output power, and must not exceed 50 mW. This means that the third harmonic of the AM transmitter must be 50 mW or less in power level. A 50-mW signal from a couple of blocks away can be mighty strong in a ham receiver! In fact, transoceanic amateur QSOs have been made at such power levels.

If any of these problems occur at your station, check for loose joints in the antenna system. If this does not resolve the difficulty, look for poor electrical joints in nearby metal objects. Once the bad joint is found, it's a simple task to clean the mating surfaces and solder them. A jumper wire and two clamps can be used to bond joints in fences and other large conductors.

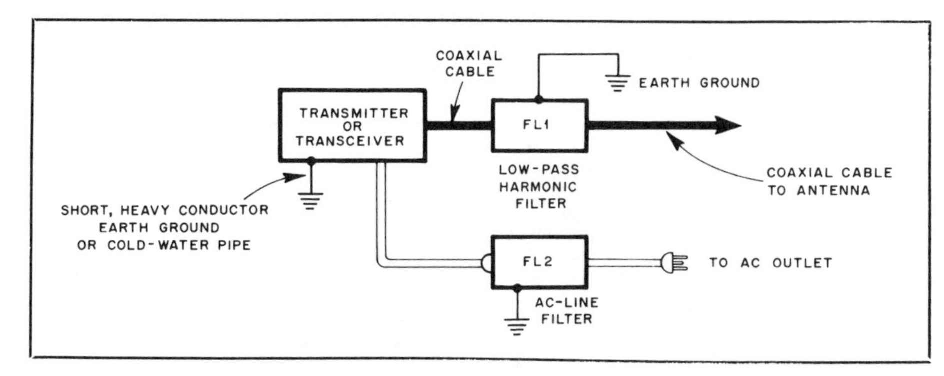

Fig. 1 — Block diagram of an amateur transmitter or transceiver that has a low-pass harmonic filter, plus a brute-force ac-line filter. The filters attenuate harmonic energy to reduce the possibility of TVI and RFI. The ground system should be of high quality (see text), with as short a connecting lead as possible. Frequently, the cold-water pipe system will serve as an effective earth-ground connection.

Interference Preventive Measures

We need to ensure that our transmitters have a clean bill of health, so to speak, before we attempt to solve interference problems in our neighbors' or our own home-entertainment equipment. Caution: If you work on a neighbor's home-entertainment device, you leave yourself open for continuing — or worsening — problems if your cure doesn't work exactly right. Fig. 1 shows the prescribed methods for keeping harmonic energy from reaching the feed line and antenna. FL1 is a low-pass filter. It allows amateur signals to pass through it with little attenuation, but frequencies above, say, 40 MHz are attenuated greatly. This filter should be located as close to the transmitter antenna jack as possible, and it should be connected to a quality earth ground. A suitable earth ground might be several 6-foot copper rods driven into the ground near the ham shack, with about 3 feet between each rod. The rods are bonded together by means of a heavy-gauge conductor, such as the shield braid from RG-8 coaxial cable. The lead from this ground system to the ham station should be as short and fat as you can make it: The shorter the overall ground lead, the more effective the ground system will be for conducting the harmonic energy to ground. Low-pass filters are widely available on the commercial market. Some amateurs build their own filters from data given in the *ARRL Handbook*.

RF energy at the operating and harmonic frequencies can be conducted along the ac line, then to the power lines for radiation. This unwanted energy may also be conducted into your neighbors' homes and then into their entertainment equipment. It should be standard practice, therefore, to install FL2, an ac-line filter. It will serve also to keep unwanted external noise and RF energy from entering your receiver via the ac line. The *Handbook* has the details for making your own ac filter. FL2 of Fig. 1 should also be located as close to the transmitter as is convenient.

A final word about harmonic radiation is in order. If you mistune the output amplifier of your transmitter (tune it to the wrong frequency), the harmonic output energy level can be quite high. Always tune your transmitter in accordance with the operating instructions. Be sure the amplifier stage is adjusted for correct loading and plate-current dip when using a tube type of output stage. This does not apply to solid-state amplifiers. They are broadband devices, and a harmonic filter is included in the circuit for each operating band.

Dealing with the Neighbor's Problem

Modern solid-state entertainment equipment is more prone to interference than was generally true of vacuum-tube equipment. This is because the transistors and ICs contain diode junctions. These diodes rectify RF energy and cause all kinds of interference problems. Also, many TV and FM receivers have front ends (tuner sections) that are not capable of rejecting non-TV or non-FM frequencies. The amateur signals enter the front end and overload them. This usually blanks out the reception entirely. Interference of this class is referred to as *fundamental overloading*. The most effective cure is the insertion of a *high-pass filter* directly at the tuner of the receiver. (As mentioned earlier, avoid working on a neighbor's home-entertainment device unless you are willing to take responsibility for the modifications you make, and for any future malfunction that could be related to those modifications.) This variety of RF filter allows the TV or FM signals to pass into the receiver, but unwanted energy below the filter frequency (1.8 to 29.7 MHz, for example) is attenuated. A high-pass interference filter (Fig. 2) will not prevent *amateur VHF and UHF* energy from reaching the front end of a TV or FM set, because the filter is necessarily designed to pass all frequencies in that range.

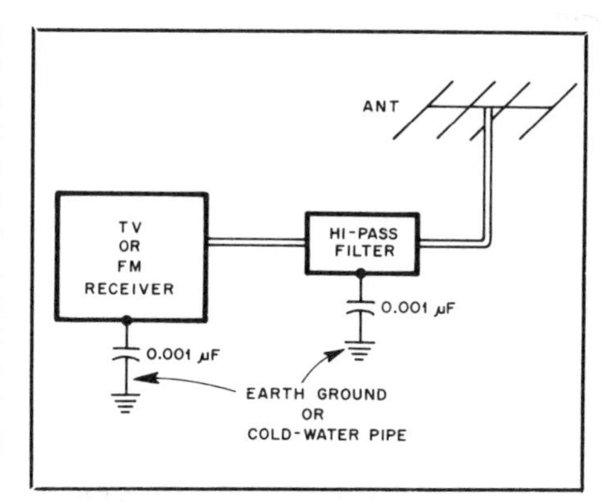

Fig. 2 — A high-pass TVI filter can be attached to the TV receiver near the tuner to prevent fundamental overloading of the TV set from amateur HF signal energy.

How FCC Regulations Affect Our Treatment of RFI-Related Problems

Q. In the eyes of the FCC, who is responsible for RFI?
A. It is the radio amateur's responsibility to ensure that the transmitted signal complies with FCC emission standards regarding the fundamental, harmonics and other transmitted products. If the transmitter is being operated in accordance with FCC regulations, the responsibility of remedying RFI susceptibility in the home-entertainment device lies with the owner of the device.
Q. What is the FCC's view on neighbor relations regarding RFI problems?
A. If a neighbor is experiencing RFI from an amateur's signal on a receiver of good engineering design, and this fact has been made known to the amateur, the FCC may impose quiet hours on the amateur. During these quiet hours, the amateur cannot transmit on the frequencies where the interference exists.
Q. What must an amateur do if faced with a notice of violation?
A. The amateur must reply to the FCC office issuing the notice within 10 days of receipt. If the notice relates to a physical or electrical problem, the amateur must state fully the steps that have been taken to correct the situation. If the notice relates to improper operation of the transmitter, the name of the operator in charge must be given. The FCC will initiate license revocation proceedings should the amateur choose not to respond.

(A)

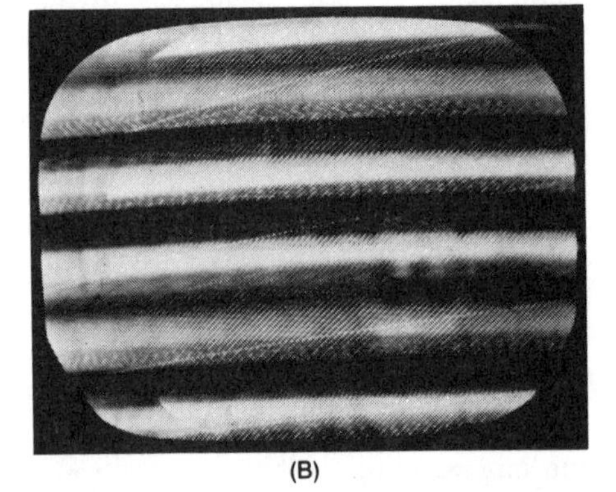

(B)

Fig. 4 — Cross-hatching is shown at A. This is typical of the harmonic interference caused by amateur transmitters. The picture at B shows what sound bars look like on a TV screen.

High-pass filters are available commercially, or you may want to make your own (less costly!) from information in the *ARRL Handbook. Caution:* Do not install any suppression device inside the neighbor's equipment. Make your installation (with his or her permission) to the equipment cabinet externally. Once you reach inside the "works," you're liable if the neighbor decides you were the cause of a subsequent equipment failure.

Fundamental overloading caused by your VHF or UHF signals must be treated in a slightly different manner. A tunable "band-elimination" filter or "band-reject" filter is generally used in the antenna lead of the TV or FM receiver. This filter is capable of rejecting your VHF or UHF signal, but passes the desired TV or FM energy to the receiver front end. This species of filter contains one or two (depending on the use of Twin Lead or coaxial feed line) tuned, high-Q circuits. They are tuned to the operating frequency of your transmitter, which will result in minimum interference to the TV or FM set. A typical circuit for this kind of filter is shown in Fig. 3. The ARRL literature covers this subject and most other items that relate to interference.[1]

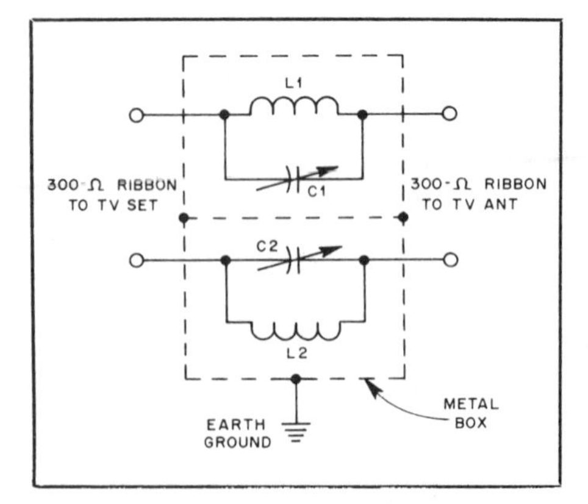

Fig. 3 — A tuned trap, or reject filter, is useful in preventing interference to FM receivers that is caused by amateur VHF or UHF energy. C1 and C2 are adjusted to resonate the traps at the transmitter output frequency.

We need to be aware that some TV set chassis are "hot" with respect to ground. This can cause an arc or even blow fuses when an earth ground is attached to the chassis. I like to stay on the safe side of things by inserting a 0.001-μF disc-ceramic capacitor in series with the ground wire to the TV receiver or filter case. This will prevent sparks from flying! (See Fig. 2.)

Harmonic Interference

Interference from amateur harmonics shows up quite differently in a TV receiver. Rather than blanking out a TV sound and picture system, harmonics cause lines on the TV screen. Diagonal or horizontal bars may appear on the screen. They may be very wide, or they may be spaced close together. Fig. 4 illustrates two kinds of "cross-hatching." Sometimes these bars appear only while you are speaking into your microphone. These are referred to as "sound bars."

It is unfortunate that we amateurs can do nothing at the TV set to cure harmonic TVI. It boils down to going back to our ham stations and starting from "square one." We must improve the harmonic suppression from our transmitters. This requires a great amount of cooperation with the neighbor while numerous checks are made to learn if progress is being made toward eliminating the problem. If the harmonic TVI occurs from operation on only one band, you should consider not using that band until you resolve your problem.

[1]C. L. Hutchinson and M. B. Kaczynski eds., *Radio Frequency Interference* (Newington: ARRL 1984).

Ac line filters are recommended for use on TV and FM sets when a tough interference problem prevails. It is possible that harmonic energy is entering the TV receiver along the ac line as well as from the antenna system. No possibility should be overlooked when trying to solve TVI or RFI problems.

Harmonic interference to FM radios must be treated in a manner similar to that for TV sets. The symptoms will show up as buzzing or voice sounds superimposed on the FM station that is tuned in. Hams who operate the 6-meter band (50 MHz) are most apt to cause second-harmonic problems to owners of FM (88-108 MHz) receivers, since the second harmonic from any transmitter is usually the strongest.

Hi-Fi Interference

Perhaps the greatest number of interference problems can be related to audio hi-fi gear. This area includes cordless telephones, electronic organs and hearing aids. For the most part, RF energy is conveyed to the equipment via the speaker leads, which are usually quite long. They act effectively as pickup antennas, thereby routing unwanted energy into the audio equipment. This difficulty is encouraged especially if the speaker wires happen to be the proper length for resonance at your operating frequency. For example, an 8-foot speaker lead would make a perfect resonant pickup antenna for 10 meters.

The most effective cure for RF energy on the speaker wires is the addition of disc ceramic bypass capacitors from each speaker terminal of the hi-fi set to chassis ground (see Fig. 5). This bypasses the RF energy to ground before it can enter the audio circuit via the back door. Another effective preventive measure is to wrap several turns of the speaker lead through a ferrite toroid core, as in Fig. 6. This acts as a choke to RF energy, but does not impair the passage of audio energy to the speakers. Once again, we should also try an ac-line filter to determine if the unwanted energy is entering the hi-fi unit along that route.

The previous methods apply to organs and other units of audio equipment. Hopefully, the required RFI-suppression components will be voluntarily included by the manufacturers in their attempts to meet the RF immunity standards envisioned by Congress when it passed PL 97-259.

Antenna Placement

It should go without saying that an amateur antenna that is close to a neighbor's house or TV antenna is a potential cause of interference. Our objective when installing an antenna should be to keep it as far from adjacent houses as possible. This is no simple assignment for the urban dweller, but physical spacing is important in preventing unwanted coupling to the nearby entertainment devices and their antennas.

Tidbits

We have not discussed interference to CATV systems. This area of difficulty can be, under some circumstances, the worst of the lot. I can recall while living in Newington, Connecticut, that I had no TVI in my own TV sets while operating the HF bands with 1 kW of power. Our TV set used an outdoor rotatable antenna. The miracle of CATV arrived in my neighborhood, and I became a subscriber. Suddenly I had TVI of the first magnitude. All efforts to cure the problem failed until I discovered that the CATV ground system was ineffective. I installed my own ground rods and solved the problem. The best approach to solving CATV difficulties is to enlist the aid of the CATV operator.

The purpose of this article is to provide you with basic information about radio-frequency interference, along with the procedures for curing RFI and TVI. The subject certainly goes much deeper than this. I recommend that you read the interference chapter in the ARRL *Handbook* and the ARRL book *Radio Frequency Interference*.

Many radio clubs have organized TVI committees. If you become the victim of poor relations with a neighbor because of interference, try enlisting the aid of a local TVI committee. It will function as a go-between for you and the irritated neighbor. Finally, don't forget that failure to attempt a peaceful solution to TVI or RFI may lead to a citation from the FCC. Good luck!

Glossary

band-elimination reject filter — a specially designed filter that rejects or suppresses a narrow band of frequencies within a wider band of desired frequencies.

decibel — a unit of relative power measurement that is used to express the ratio between two levels of power. It is equal to 10 times the common logarithm of this ratio. The abbreviation for decibel is dB. The abbreviation dBW is referenced to a power level of 1 W. The term dBm follows the same rule, but is referenced to a milliwatt (mW) rather than a watt (W).

fundamental overloading — the unwanted blanking out of the picture and sound of a TV set, caused by large amounts of RF energy from a nearby transmitter fundamental output signal. This condition is not related to harmonic energy from the transmitter, unless the offending harmonic is unusually strong.

high-pass filter — a filter designed to pass all frequencies above a desired one, while rejecting those that lie below the filter-design cutoff frequency.

low-pass filter — a filter designed to reject all frequencies above a desired one, while passing all below the filter-design cutoff frequency.

peak output — RF output power that is averaged over a carrier cycle at the maximum amplitude that can occur with any combination of signals that may be transmitted. More simply, the maximum instantaneous power output from the transmitter.

QRP — from the international Q code meaning "Shall I reduce power?" Also, "Will you reduce power?" This term is commonly used to denote amateur transmitter power levels at or below 10-W dc input to the last transmitter stage or 5-W output. Many hams are QRP operators by choice for the purpose of meeting the challenge of working long distances with very low power.

RFI — radio-frequency interference. Interference to AM and FM radios as well as to various appliances, such as computers, audio systems and telephones.

TVI — interference to television receivers.

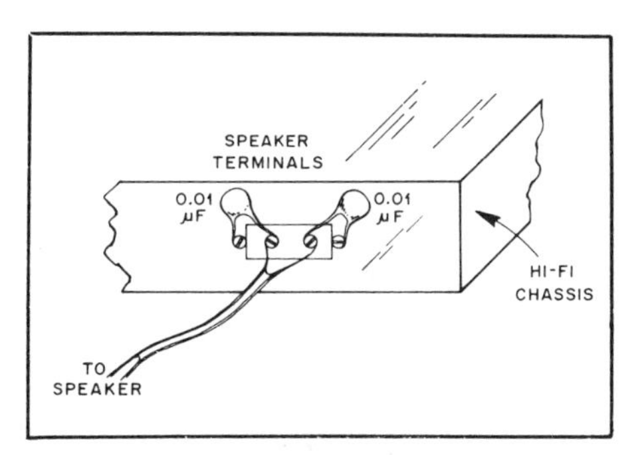

Fig. 5 — Method for reducing unwanted RF pickup by the speaker wires of an audio amplifier. A 0.01-µF ceramic capacitor is connected from each speaker terminal to chassis ground, as shown.

Fig. 6 — Winding the speaker wires on a toroid core can prevent RF energy from entering the circuit of an audio amplifier via the speaker leads. An Amidon FT-140-61 core is suitable if six to eight turns of speaker lead are looped through it.

Diodes and How They Are Used

Part 15: However simple diodes may be, you'll use them again and again in Amateur Radio.

Could there be an electronics technology without diodes? Probably not — at least not as we know the technology today. It is incredible that so simple a device can play such an important role in radio circuits. The first radio signals were detected by means of diodes: Diodes were used in "crystal sets" to detect standard AM broadcast signals in the early days of radio. There are many new types of diodes, but a diode remains a diode with respect to the function. A diode is a device that has a cathode and an anode, and passes current in one direction only.

The combination of the galena crystal and "catwhisker," used to detect AM (amplitude modulated) radio signals, was, in fact, a solid-state diode. The galena serves as one half of the diode, and the tip of the catwhisker (fine wire) comprises the remaining half. When the two objects are in contact with one another, it is possible to rectify the incoming radio signal and create pulsating dc that would activate a pair of earphones at an audio rate, thereby enabling a person to listen to a favorite radio program. A number of other materials, such as carborundum, were used to form a diode for signal detection. The idea is to provide a poorly conducting junction that causes rectification (changing ac to dc) of the radio signals.

The irony of having solid-state diodes in the old days is that vacuum-tube diodes were used for nearly every other diode application until copper-oxide and selenium rectifiers were introduced, prior to World War II. The large-signal semiconductor diode (silicon or germanium) came into being in the early 1950s. Germanium small-signal (low-power) diodes were used prior to 1940 for various detector circuits, and were a vital part of radar receivers during WWII.

Modern Diodes

Fig. 1 illustrates the progression of diodes since the first days of radio. We went from the tube diode and galena crystal to selenium diodes, low-power point-contact (germanium) diodes and, finally, to germanium or silicon junction diodes. Germanium has fallen out of popularity as a power-diode material; silicon is the principal material used today. We now have rectifier diodes that can accommodate many amperes at relatively high peak voltages (e.g., 50 A at 100 V). The larger diodes are used extensively in such devices as electroplating rectifiers, welding machines and automobile alternators.

The small-signal diode has come a long way, also. There are many types of internal structures for these diodes, and each is created to perform a specific job in electronic circuits. We will examine some of these interesting applications later.

One of the principal differences between germanium and silicon diodes (apart from the difference in crystal material) is the "barrier voltage." If we apply a current to a diode, it will not commence to conduct until a particular voltage is developed across the diode junction. Therefore, this level of voltage functions as a *barrier*. For germanium diodes, the barrier voltage is approximately 0.4. It is on the order of

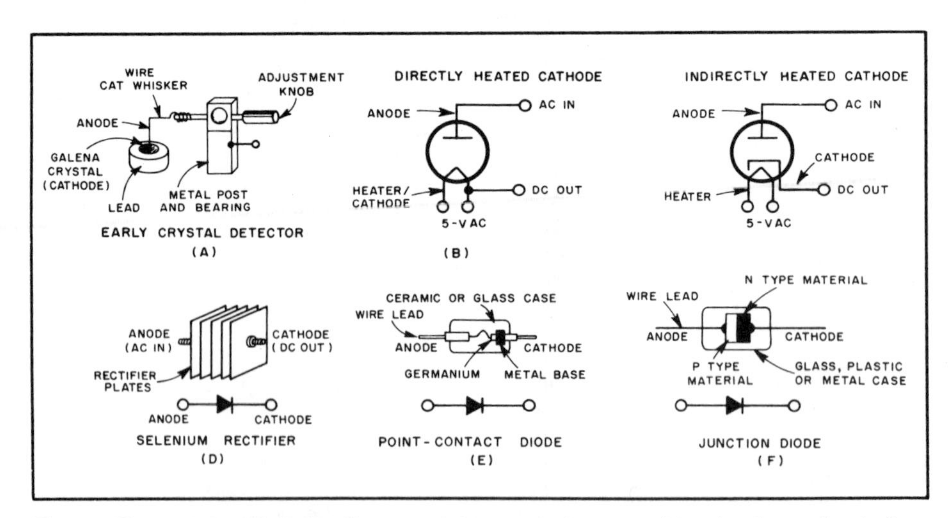

Fig. 1 — The evolution of diodes. The crystal detector at A was used for signal reception in the early days of radio. Vacuum-tube diodes (B and C) were used as rectifiers and signal detectors for many years, until selenium (D) and junction diodes (F) replaced them. Point-contact diodes remain in common use as hot-carrier diodes (E).

0.7 V for a silicon diode. In everyday language, we may think of the germanium diode as being the more *sensitive* of the two, since it will conduct at a lower voltage level than will its silicon brother.

The solar-electric (photovoltaic) cell found in solar panels utilizes the barrier concept to develop dc voltage. A solar cell is a type of diode, and when photons impinge on the cell a current will flow. The barrier voltage is approximately 0.5 per cell. Therefore, many solar cells must be wired in series to obtain a desired voltage output from a solar panel. To realize 13 V from a solar panel (with a load connected to it, such as a transmitter or receiver), we must use roughly 36 cells. Under no load this will produce about 17-18 V dc at peak sunlight.

Various packaging formats are used for the current crop of diodes. They may be housed in metal, glass or plastic cases. Fig. 2 shows a group of modern diodes in various packages. The physical characteristics may vary with the power ratings, and electrically similar diodes may look different because of the manufacturer's choice of package style. Most diodes have a band of paint around one end of the body to identify the cathode end of the diode. The cathode end of a metal-encased, stud-mount power diode is generally the stud end. The anode terminal is set in glass at the opposite end of the diode.

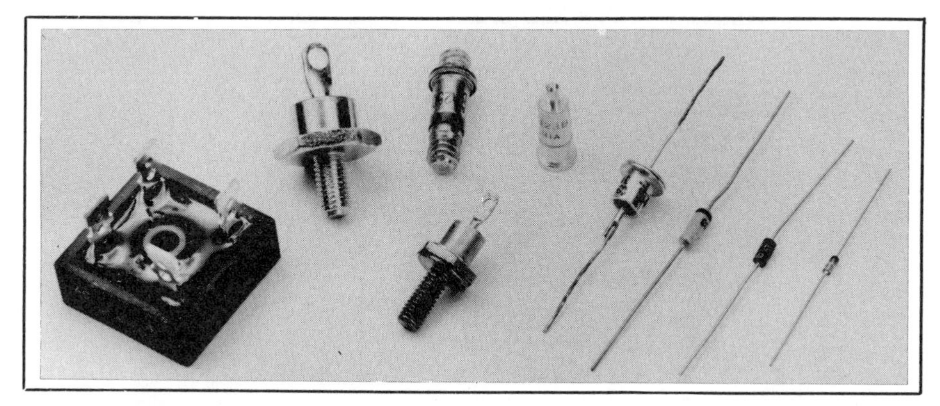

Fig. 2 — Various types of solid-state diodes. High-power units appear at the left, with small-signal diodes at the right.

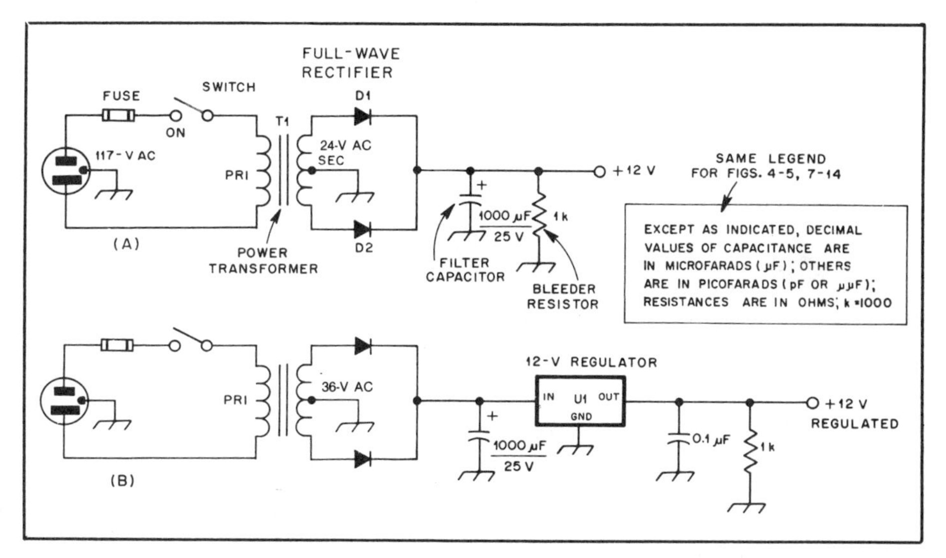

Fig. 3 — Diagram A shows a simple, unregulated 12-V dc power supply in which D1 and D2 serve as rectifiers. A regulator (U1) has been added in circuit B to stabilize the output voltage at +12.

Diodes as Power Rectifiers

Whenever we apply ac voltage to a diode and extract dc voltage from it, we are using the diode as a *rectifier*. Even though the diode may be used in a specific application to change a radio-frequency voltage (signal) to audible, pulsating dc voltage, it is still acting as a rectifier.

Few pieces of electronic equipment operate without some form of ac power supply. Among the exceptions are portable radios, watches, calculators, car radios, and some portable and mobile Amateur Radio gear. Most indoor appliances are plugged into the ac outlets of our homes. When this is done, we must have provisions, within or outside the equipment, to not only increase or decrease the wall-outlet voltage to a level suitable for the item it will power, but to change the ac voltage to dc.

As we learned earlier in this series, a transformer is used to change the voltage amount. Let's suppose that we wanted to power a CW keyer from the ac wall outlet. The keyer is designed to operate from 12-V dc. What type of power supply would be suitable to satisfy our requirements? A circuit for accomplishing our goal is shown in Fig. 3A. However, we will find that the dc voltage will shift up and down somewhat as the CW keyer is activated (no load to full load). Some circuits are not sensitive to small voltage changes, while others are very intolerant of voltage shifts (poor regulation). The no-load, full-load voltage changes can be stopped if we add a voltage regulator, as shown in Fig. 3B. I don't wish to get involved with a discussion of voltage regulators in this installment, but you should be aware that they exist and that they are used frequently to ensure a nearly constant output voltage from a power supply.

The power supply at A of Fig. 3 will produce approximately 17-V dc while no external load is attached to it. The output voltage will drop to roughly 12 with the load connected. D1 and D2 are the rectifier diodes that change the ac to dc.

A voltage regulator has been added to our power supply, as shown at Fig. 3B. Another change is the power-transformer secondary voltage. It has been increased to 36 V. This is necessary in order to permit U1 to work as a regulator: A regulator needs more input voltage than the output voltage it delivers. Actually, this circuit delivers 25.3-V dc to the input side of the regulator. This is because when a filter capacitor is used immediately after a rectifier, the no-load voltage from the diodes is 1.41 times half the secondary ac voltage of the power transformer when a full-wave rectifier is used. The input voltage to a regulator must be high for another reason: For the regulator to prevent the power supply output voltage from falling *below* the desired amount, more than the required output voltage must be present in the first place (25.3 V versus 12 V).

Rectifier diodes can be used in other forms of power supplies. For example, we may use what is called a half-wave rectifier (Fig. 4A), but the load-no-load shift in voltage will be greater than with a full-wave rectifier. Also, the ripple (hum) from a half-wave rectifier is much harder to filter out. Fig. 4B shows a full-wave bridge rectifier. The dc voltage characteristics from this rectifier are the same as for the simple full-wave rectifier of Fig. 3. When the four diodes are used, however, the power transformer does not require a center tap on the secondary winding.

We can employ diodes in other types of

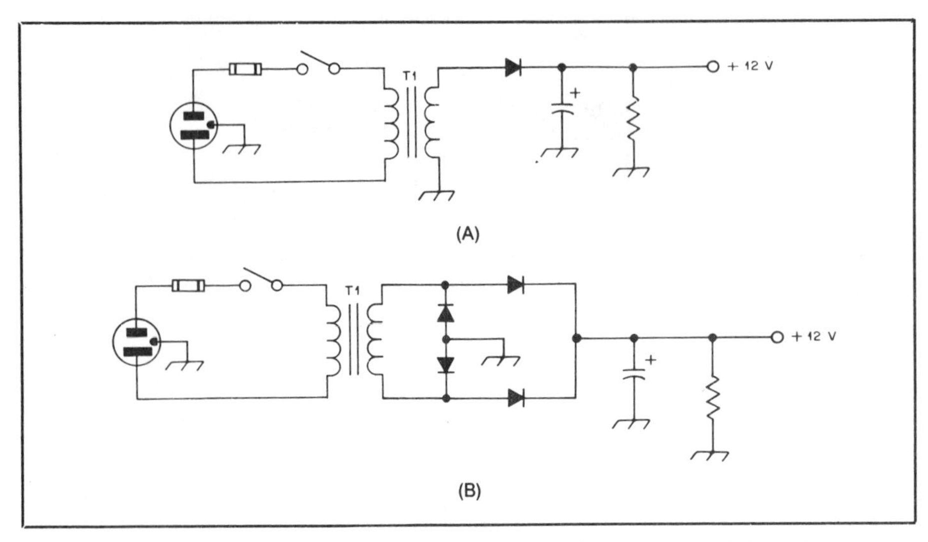

Fig. 4 — Diodes are shown here as rectifiers in a half-wave (A) and full-wave bridge (B) rectifier. T1 needs no secondary center tap when a bridge rectifier is used.

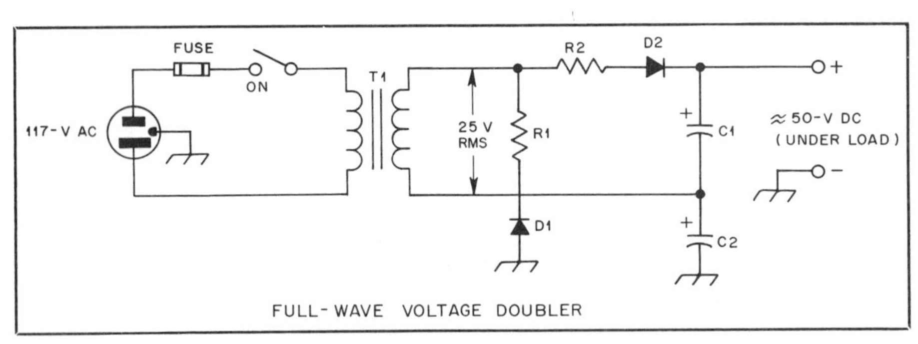

Fig. 5 — Example of how two power diodes can be used in a voltage doubler.

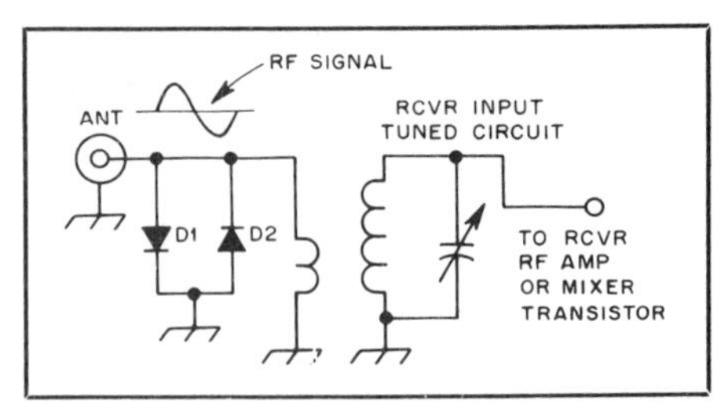

Fig. 7 — Small-signal diodes can be used as shown to prevent damage to the front end of a receiver from excessive input signal voltage.

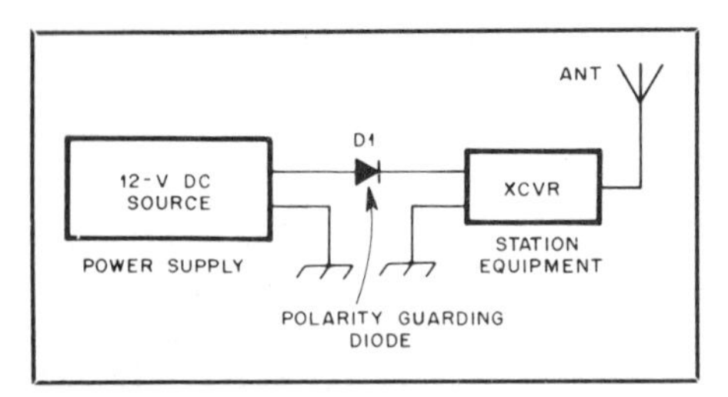

Fig. 6 — D1 functions here as a protective diode against accidental polarity reversal of the power supply (see text).

power-supply rectifiers. They can be arranged with appropriate capacitors to form a voltage doubler, as shown in Fig. 5. The resistors (R1 and R2) are chosen to protect the diodes from high-current surges when the power supply is turned on. These resistors will also protect D1 and D2 from excessive reverse voltage: Too much voltage or current can destroy a diode.

The dc output from a voltage doubler of this type will be approximately twice the RMS voltage across the secondary winding of T1. Under no-load conditions the voltage may approach 2.7 times the RMS secondary voltage.

There are also voltage triplers and quadruplers. A full explanation of power-supply rectifiers and their applications can be found in the power-supply chapter of the ARRL *Handbook*. I urge you to go beyond this simple discussion of diodes by studying the *Handbook*.

Other Uses for Diodes

A diode can be used as a protective gate. An example of this is shown in Fig. 6. D1 is inserted between the dc power supply and the equipment with which it will be used. If we were to reverse the power-supply terminals (reverse polarity), we could instantly destroy the solid-state devices in our transceiver or other equipment. Such mistakes are made frequently. We may prevent damage resulting from human error by using D1 of Fig. 6. It will permit current to flow through it when the power-supply polarity is correct. Current will not pass through D1 if the polarity is reversed, thereby protecting the station equipment. There will be a 0.7-V drop through D1 (the barrier voltage), so the power supply should have an output of 12.7 or 13 V to ensure that 12 V reaches the equipment. The diode must be chosen to safely pass the current of the transceiver or other gear with which it is used. Similarly, we must select a diode that has a voltage rating somewhat greater than 12 V for this example.

Another protective circuit in which diodes can be used is found in Fig. 7. Here we show two diodes reverse-connected in parallel across the 50-ohm receiver input line. They will not conduct until the incoming signal voltage (ac) reaches approximately 0.7. They will create a short circuit for all input signal voltages in excess of 0.7. D1 and D2 will prevent damage to the input stage of the receiver. Arranged as shown in Fig. 7, D1 and D2 will conduct on both the negative and positive peaks of the incoming RF signal. We may wish to use two diodes in series for each leg of the protective circuit. The barrier voltage will then be 1.4, which will still ensure safety for the receiver. Series diodes are sometimes necessary when very strong commercial signals are present. They could cause D1 and D2 of Fig. 6 to conduct, which would result in rectification of unwanted signals. This would create many spurious signals and "hash" to appear in the receiver output. By using two diodes in series for each branch of the protective circuit, we would raise the barrier voltage above the signal level of the strong commercial station.

Fig. 8 shows how we can use a silicon power diode to establish a 0.7-V positive potential that is used as bias for a solid-state linear amplifier. R1 is used to limit the current through D1, thereby preventing the diode from burning out from excessive heat. Here again, we have taken advantage of the barrier voltage of the diode to establish a +0.7-V reference for the base of Q2.

Diodes are commonly used as electronic switches. We can see how this is done by referring to Fig. 9. The advantage is that S1 can be located a long distance from the three crystals, Y1, Y2 and Y3. The leads going to the diodes carry only dc voltage. The three 4.7-kΩ resistors provide RF isolation between S1 and the crystals, while serving as current-limiting devices for the diodes. As each diode is made to conduct, via application of dc voltage from S1, the related crystal is connected to the oscillator circuit. Low-power silicon diodes of the high-speed switching variety are suitable for this type of circuit. This same general switching technique is used for selecting

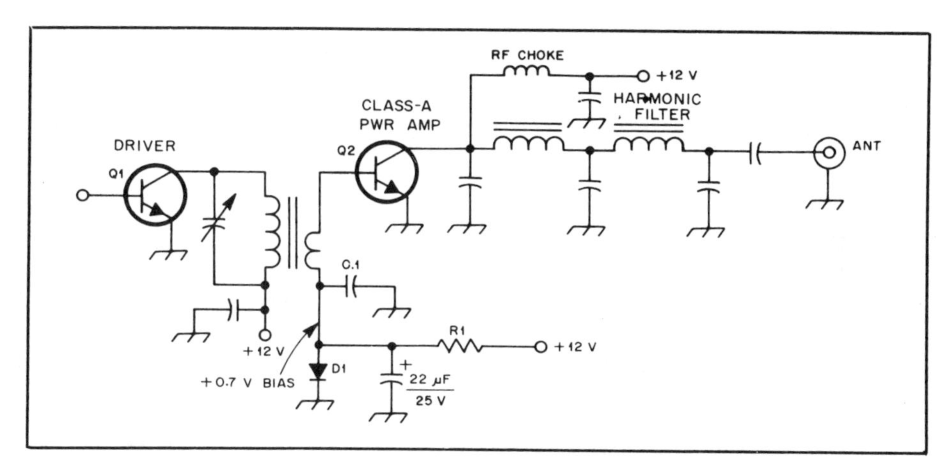

Fig. 8 — The barrier voltage of a silicon diode can be used to establish a +0.7-V bias for linear operation of an RF power transistor.

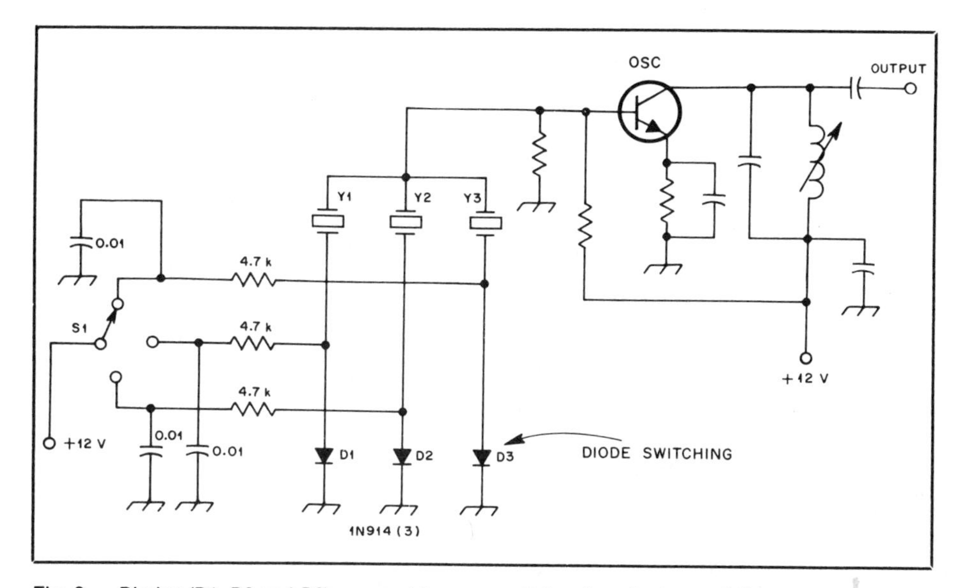

Fig. 9 — Diodes (D1, D2 and D3) are used here as switches to select one of three crystals.

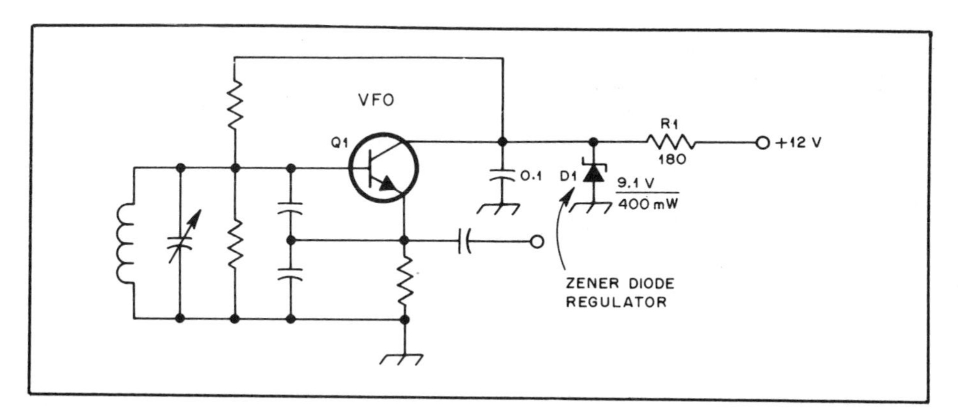

Fig. 10 — Zener diodes serve as regulators when connected as shown. They are available in a host of voltages and power ratings.

various tuned circuits and filters in radio equipment.

Still More Applications

Applications for diodes are limited only by your imagination. We are barely touching the surface in this article, but let's examine a few more common uses for diodes. A special type of diode is shown in Fig. 10. D1 is known as a Zener diode, and you can observe that the symbol has two little hooks on the cathode end. Only a Zener diode has this symbol. We see D1 serving as a voltage regulator for a VFO. R1 is a current-limiting resistor that prevents D1 from passing to much current through its junction. The formula for selecting the correct R1 values is given in the *ARRL Handbook*. In this circuit, D1 maintains the Q1 operating voltage at 9.1, despite variations in the 12-V supply line. Severe voltage changes would cause the VFO to change frequency unexpectedly, so we have used a Zener diode to stabilize the dc voltage at Q1. Zener diodes are available in various operating voltages and power ratings.

In Fig. 11 we see a pair of low-power, high-speed diodes employed in a frequency doubler. You will notice a similarity between the hookup for T1, D1 and D2 and the circuit of Fig. 3A. A doubling action takes place also in the full-wave rectifier of a power supply. That is why the frequency changes from 60 to 120 Hz in a power supply.

Another style of diode is called a varactor (variable reactor). This diode is illustrated in Fig. 12. It can be used as a frequency doubler, tripler and quadrupler, or to generate higher-order harmonics of the driving signal. Here we depict it as a tripler. The second harmonic is removed by what is called an "idler tank," consisting of C1 and L1. Varactor diodes are quite efficient. For example, if we fed 25 W of 144-MHz energy into J1, we could obtain as much as 17.5 W of output at 432 MHz (J2). No dc operating voltage is needed.

A varactor type of diode can be used as a tuning diode. Fig. 13 contains two examples of how this is done. A single tuning diode is shown at A of Fig. 13. As the positive voltage applied to the cathode of D1 is changed by means of R1, the internal capacitance of the diode changes, thereby tuning L1 to resonance at various frequencies. Many modern TV sets use tuning diodes in the front-end section to avoid the use of a mechanical channel

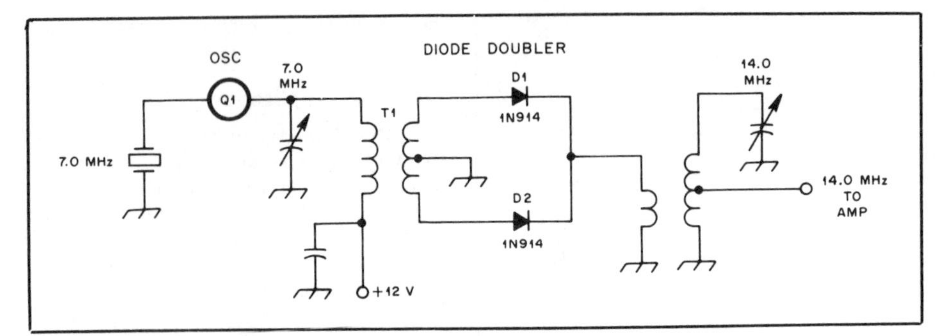

Fig. 11 — A push-push type of frequency doubler is shown here. D1 and D2 provide output at the second harmonic of the driving signal.

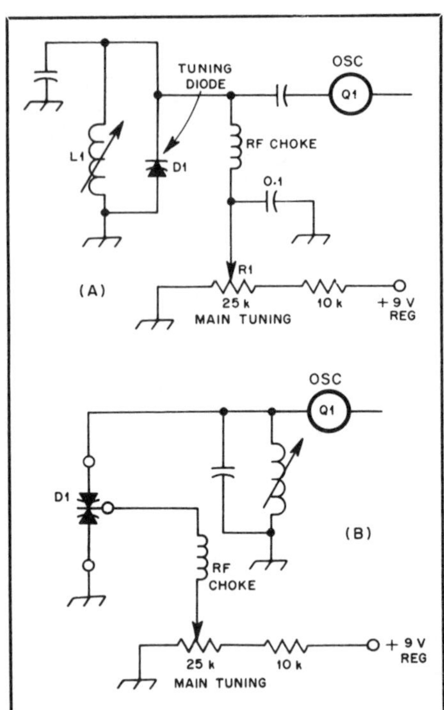

Fig. 13 — Tuning diodes can be used in place of large variable capacitors (mechanical). A single-ended tuning diode is illustrated at A, while the preferred type is shown at B (D1) in a double-ended format. The internal capacitance of the diode changes as the applied operating voltage is raised or lowered by means of R1.

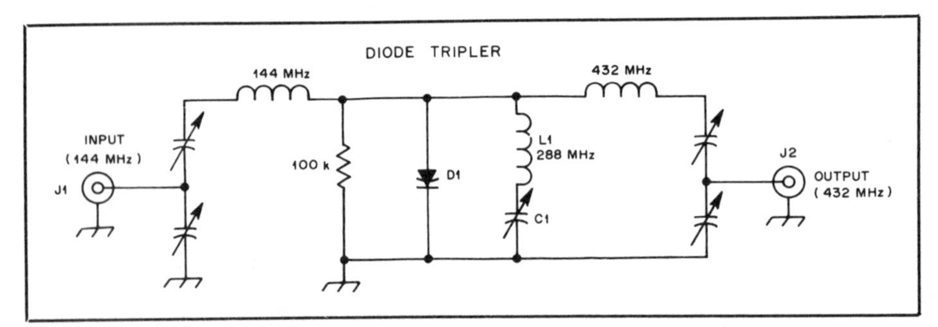

Fig. 12 — A power frequency multiplier can be made from a varactor diode. This circuit is typical of that used for tripling from VHF to UHF.

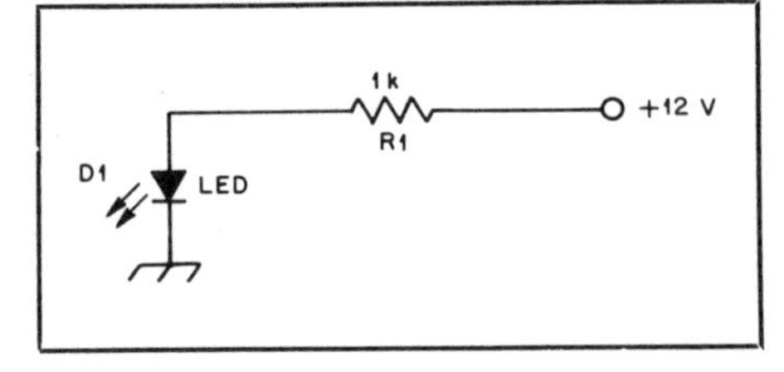

Fig. 14 — Example of an LED with operating voltage applied.

Glossary

barrier voltage — the threshold voltage required across a diode junction to make it conduct or turn on. For germanium diodes it is roughly 0.4 V, while for silicon diodes it is 0.7 V.
full-wave rectifier — a rectifier in which the negative half of the sine-wave cycle is inverted so that the output contains two half-sine pulses for each input cycle.
half-wave rectifier — a rectifying circuit that passes only one half of the incoming sine wave, and does not pass the opposite half cycle. The output contains a single half-sine pulse for each input cycle.
LED — light-emitting diode. It illuminates when operating voltage is applied: positive to the anode, negative to the cathode.
linear amplifier — an amplifier for which the output waveform is a faithful reproduction of the input waveform, or the output quantity is essentially proportionate to the input quantity.
photovoltaic — a principle in which a photoconductive or photoemissive action takes place. Transparent conducting films are separated by semiconductor material to form a photovoltaic cell (solar cell). Electromagnetic radiation upon one of the films will create a potential difference between the films.
rectifier — a device that converts alternating current (ac) to direct current (dc). It has the characteristic of conducting current substantially in one direction only.
varactor diode — a diode for which the internal capacitance is voltage-dependent. Used for frequency multiplication and electrical tuning of LC circuits.
Zener diode — a special diode used for voltage regulation. A diode that exhibits, in the avalanche-breakdown region, a large change in reverse current over a very narrow range of reverse voltage.

switch in the critical RF circuits.

A double tuning diode is shown in Fig. 13B. It is the preferred type of varactor in terms of linearity of the tuned circuit. One such diode is the Motorola MV104. Various capacitance ranges are available for tuning diodes. This means that we must choose the proper diode for the desired tuning range.

We must not neglect to mention a very familiar modern-day diode — the LED. LED means "light-emitting diode." They are available in many colors. When the LED is given enough current to make it conduct, it illuminates. A circuit for 12-V use is illustrated in Fig. 14. R1 is a current-limiting resistor that prevents burnout of D1. LEDs have a cathode and anode, just as do the other diodes. For this reason they will not light if the wrong polarity of voltage is applied. The electrical symbol for an LED always has two arrows pointing away from the cathode, as shown in Fig. 14.

Summary

An entire volume could be written about each of the diodes we have examined here. Our objective has been to familiarize you with some of the more common uses for diodes. The list of practical applications in amateur circuits goes on and on; as you continue to gather knowledge, you will become familiar with all manner of uses for diodes. I dare say you will think of some applications that have escaped me! Once again I want to urge you to pick up your *ARRL Handbook,* and study the chapter on semiconductors and power supplies. The real nitty-gritty of how diodes function and how you may use them is contained in those chapters. It will be helpful for you to wire up some simple experimental circuits that use diodes. Observing the action of diodes will help you to better understand them.

Resonance and Tuning Methods

Part 16: Have you ever wondered how a piece of radio gear is able to tune in a particular frequency?

The beginner series can't be explicit in all areas. No matter where we go to learn new subject matter, we should always consider additional reference material. All the articles in this book can be greatly enhanced if you are willing to dig deeper into the subjects covered. Don't be reluctant to do so![1]

Tuned Circuits

Radios contain many tuned circuits (made up of coils, also called inductors, and capacitors) that are set for a particular frequency. These take many forms. Furthermore, a given radio receiver or transmitter will have circuits that are tuned to many frequencies. Seldom are all of the tuned circuits adjusted for operation on the same frequency.

An understanding of tuned radio-frequency circuits is helpful in your quest for electronics knowledge. When you graduate to the level of home-equipment design and repair, it is essential that you know how these circuits operate. Another term you will encounter is "resonator." This is simply a coil and capacitor combination that forms a tuned circuit for a selected frequency. The inductors in these circuits are also known as "tank coils," especially in transmitters. You will hear amateurs mention the "final amplifier tank," for example. This refers to the coil and capacitor used in the output tuned circuit of the amplifier.

Still another popular expression for a coil-capacitor combination is "network." For example, the matching network between a transmitter driver stage and the power amplifier may contain coils and capacitors that are used to change one impedance to another, such as a 50-ohm driver output to a 10-ohm power-amplifier input line. As we learned earlier in the series, maximum power transfer will take place only if unlike impedances are matched.

Like resistance, impedance is expressed in ohms. An impedance usually consists of resistance and reactance. The symbol for impedance is the letter Z. If an impedance happens to contain only resistance, it is defined as "resistive impedance."

Since we have mentioned reactance, it would be wise to discuss it here. Reactance is also expressed in ohms. There are two kinds — capacitive and inductive. The impedance of a capacitor or an inductor changes as the frequency of operation changes. The part of the impedance that is frequency dependent is called "capacitive reactance" in capacitors and "inductive reactance" in inductors.

Capacitive reactance is expressed as X_C and inductive reactance is expressed as X_L. When a coil and capacitor are said to be tuned to resonance, the inductive and capacitive reactances are equal but opposite in action. When this condition is met, the reactances cancel one another. The tuned circuit then looks like a pure resistance at the frequency of resonance, a desirable condition. A more detailed explanation of reactance and impedance can be found in *Understanding Amateur Radio*

Fig. 1 — A tunable coil-capacitor combination. This is a parallel-resonant circuit. As the capacitor is adjusted through its range, various frequencies of resonance are established. The coil inductance and capacitance range are chosen to cover a specific range of frequencies.

[1]Several League publications are designed for this purpose. The first is *Tune in the World With Ham Radio*, which contains a clear, basic explanation of radio theory. If you'd like more detail, *Understanding Amateur Radio* is a good source. The best all-around reference for Amateur Radio operators, of course, is the *ARRL Handbook.*

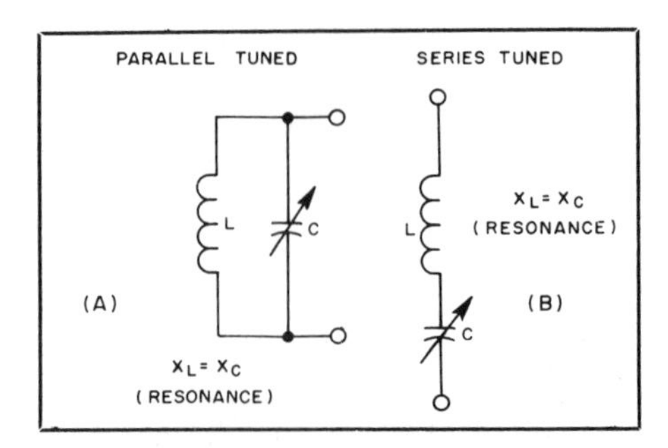

Fig. 2 — Illustration A is the electrical equivalent of the tunable coil capacitor in Fig. 1. It is a parallel-resonant circuit. A series-resonant circuit is shown at B.

or the *ARRL Handbook*.

Conventional Tuned Circuits

Let's look at some older types of tuned circuits to see how they are hooked up. Fig. 1 shows an air-wound coil and a mechanical tuning capacitor. The inductance value of the coil is not changed in this arrangement. Rather, the capacitor is adjusted to change the resonant frequency of the coil-capacitor combination. At each setting of the capacitor, we will have resonance (canceled reactance) at a different frequency within the adjustment range of the capacitor. Because of this ability to change the resonant frequency, the variable capacitor is called a "tuning capacitor" or a "resonating capacitor."

The electrical equivalent of the parts in Fig. 1 is offered in Fig. 2. The arrow through the capacitor indicates it is adjustable. The illustration at A represents what is known as a parallel-resonant circuit, since the coil and capacitor are in parallel. At B of Fig. 2, the same parts are arranged in a series-resonant circuit. In both instances, the reactances must cancel one another for the circuit to resonate.

Tuning capacitors take a host of forms. Some are adjusted by means of screwdrivers or tuning tools. These are generally called trimmers or padders. They are set for resonance just once, then left in that position. Trimmer capacitors may be made with metal plates; insulation between the plates can be made of ceramic, plastic, mica or glass. One type has a movable conductive plunger that is adjusted inside a glass cylinder that has a conductive outer coating at one end. These are called "piston trimmers."

Fig. 3 shows various trimmer and padder capacitors that might be used in a tuned circuit. Some are more desirable than others, since they are mechanically superior to the less-expensive types. This helps to ensure that they remain set to a specific capacitance in the presence of vibration or temperature changes. The insulation used between the movable plates of a trimmer capacitor also affects the performance. The better the dielectric quality of the material, the better the capacitor for RF tuned circuits. Some trimmer capacitors resemble the larger variable capacitors, except that they are miniature versions of the larger units. Air is the insulating material between the plates.

Other Tuning Methods

A coil-capacitor combination can be used to cover a range of frequencies by using a fixed-value capacitor and a variable coil (Fig. 4). You will see many such circuits in pocket-size AM and FM broadcast receivers. These are little metal cans with tuning slugs that are accessible through holes in the cans. The resonant frequency is changed by adjusting the coil slug. The slugs are made from powdered iron or ferrite, which increases the coil inductance as it is moved farther into the coil. Brass slugs can be used in place of the powdered-iron ones, but the farther they are inserted into the coil the smaller the inductance becomes — the opposite effect from powdered iron or ferrite. Fig. 5 contains the electrical details of a number of adjustable inductors. You can see that various mechanical schemes make it possible to change the effective inductance of a coil.

Modern Tuned Circuits

Circuits used in modern radio gear are tuned electronically rather than mechanically. A semiconductor, such as a diode or transistor, serves as the tuning capacitor. Most modern TV receivers are tuned in this manner (varactor tuning).

An example of a typical electronically tuned circuit is given in Fig. 6. D1 is a special diode manufactured for use in tuning an inductor to resonance or for changing the oscillation frequency of a quartz crystal. Il-

Fig. 3 — A collection of trimmer and padder capacitors. These units have variable capacitance, but are not suitable for use as a main-tuning control because they have no shafts on which to mount a knob or dial mechanism.

Fig. 4 — A slug-tuned coil (foreground) and a roller inductor coil. These inductors are variable, which permits the use of a fixed-value capacitor to form a tunable resonant circuit.

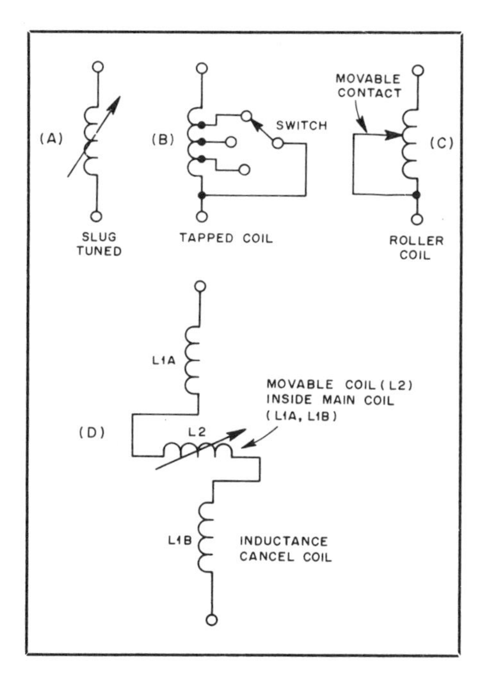

Fig. 5 — Electrical symbols for a slug-tuned coil (A), a tapped coil (B), a roller inductor (C) and a bucking or canceling coil (D). L2 of D rotates inside L1 to aid or oppose the total inductance, thereby changing the effective inductance. This old-style coil is seldom used.

lustration A is for a VXO (variable-frequency crystal oscillator), while drawing B demonstrates schematically the arrangement for a resonant coil-capacitor circuit that uses a tuning diode. In both circuits, the internal junction capacitance of the diode is varied by changing the dc voltage applied to the diode. This is done at R1, which can be mounted on the equipment panel for accessibility. For the most part, the outward appearance of a tuning diode is no different from that of a rectifier diode or glass Zener diode.

Is there a shortcoming to the use of tuning diodes? Yes. They are not as frequency-stable as are mechanical variable capacitors: The diode junctions are affected by changes in temperature; therefore, they will change capacitance gradually after the operating voltage is applied. This causes frequency drift. Also, the ambient-temperature changes within a piece of equipment will cause the diode capacitance to vary slightly — a further cause of drift. Of course, mechanical capacitors are also affected to some extent by temperature changes, but not so dramatically as are tuning diodes. For many circuits, however, it is entirely practical to use tuning diodes.

Another important difference between mechanical and electronic tuning capacitors is the minimum-to-maximum capacitance range. Take, for example, a 100-pF (maximum capacitance) air-variable capacitor. When the plates are fully meshed it will provide 100 pF. With the plates completely unmeshed the capacitance might be only 10 pF. A tuning diode, on the other hand, may have a maximum capacitance of, say, 100 pF, but the minimum capacitance may be 35 pF. This means that the tuning range with a given coil will be smaller than that with a mechanical tuning capacitor.

Once we recognize and accept the peculiarities of tuning diodes, we can proceed to use them in our amateur circuits. They are far less expensive to use than air-variable capacitors and they permit miniaturization that would otherwise be impossible. A tuning diode can cost as little as 35 cents, whereas a mechanical equivalent capacitor could cost $20!

Tuned Circuits Versus Power

We have thus far overlooked the matter of RF and dc power that must be accommodated by a tuned circuit. In transistor or IC circuits, we need not be too concerned about the operating voltages of the variable capacitors we use. It is unusual to have more than 28-V dc in a semiconductor circuit, so small capacitors can be used without fear of arcing or overheating. This is not true in circuits that use vacuum tubes — particularly at high RF and dc voltage levels. The greater the voltage, the wider the plate spacing of the variable capacitor must be to prevent voltage from arcing across the capacitor plates. In circuits where high RF power is developed, it is wise not to allow dc voltage to be applied to the tuning capacitor, even though some amateurs have done this in homemade gear. Fig. 7 shows both methods, but the illustration at B is recommended for safety reasons as well as voltage-breakdown considerations.

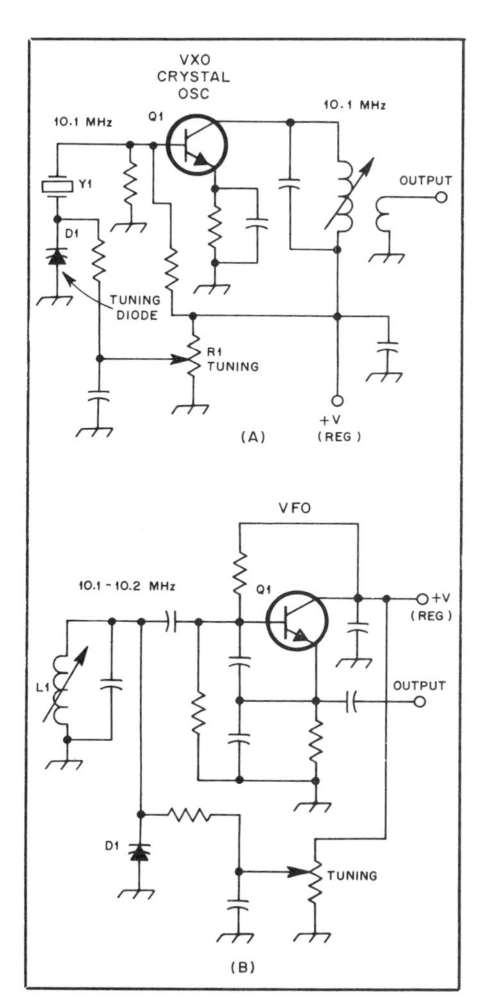

Fig. 6 — A crystal (Y1) can be shifted in frequency by using a tuning diode (D1) in series with it. As the dc voltage applied to D1 is varied by means of R1, the diode capacitance changes, and this changes the oscillation frequency of Y1 slightly, as indicated at A. A VFO can be tuned (B) by using a tuning diode in place of a mechanical tuning capacitor.

Glossary

dielectric — an insulating material or medium that, ideally, has zero conductivity. A vacuum is a perfect dielectric. Solid materials, such as glass or ceramic, are suitable insulating materials, but not as good as air or a vacuum.

drift — a condition under which an electrical property changes as a function of temperature — internal or external to the components. Frequency drift is associated with oscillator circuits that change frequency as a function of time and heating.

network — a group of conductive components connected to form a specific circuit function. A group of coils and capacitors can, for instance, comprise a network.

resonator — a device intended to introduce resonance into a circuit. A coil-capacitor combination tuned to a specific frequency is a resonator. A quartz crystal is also a resonator.

tuning diode — a special type of semiconductor diode that exhibits a significant change in junction capacitance as the dc potential across the junction is varied. A varactor is a type of tuning diode.

The power-handling capability of a coil is dependent on the wire size. The greater the dc current or RF current that flows through a coil, the larger the conductor must be. The smaller wire has a higher dc resistance in ohms. The ac (or RF) resistance is higher also. Ohm's Law tells us that the greater the resistance and current in a conductor, the higher the loss, or voltage drop. Power is dissipated in the wire, and dissipation causes heat. A coil with too small a wire size can burn up easily. The coil resistance should always be as low as possible. The insulating material on which the coil is wound should also be able to sustain the developed or operating voltage without burning or arcing. If the wire in a coil becomes more than slightly warm after a few moments of circuit operation, chances are that the wire diameter is too small. Heating losses impair the efficiency of a circuit.

A Laboratory Experiment

Perhaps you have a desire to warm up a soldering iron and try your hand at building an electronically tuned circuit. This would be a fine way to observe what happens when the voltage on a tuning diode is varied. You will need a shortwave radio or ham receiver for this lab experiment. You will be building a 40-meter oscillator, and you will want to listen to the signal in a receiver.

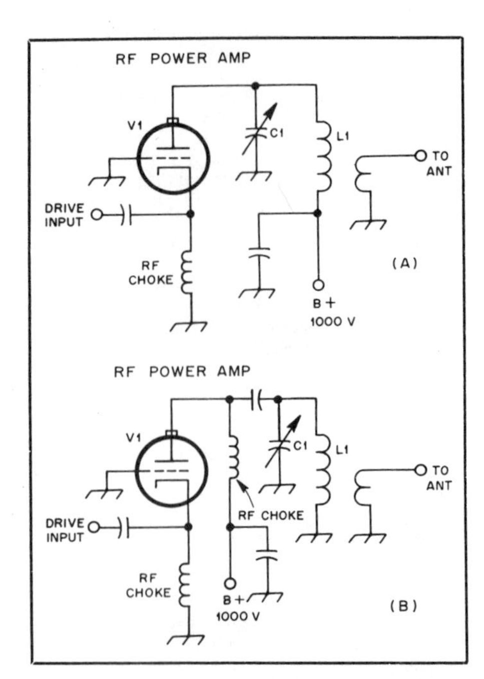

Fig. 7 — High voltage should be kept off the tuning capacitor (see text). Illustration A shows series feed of the B+ voltage (poor practice), whereas shunt or parallel feed is depicted at B (desirable).

Fig. 8 is the schematic diagram of our voltage-controlled VFO (variable-frequency oscillator). An MPF102 or 2N4416 FET functions as the oscillator transistor. Any high-frequency N-channel junction FET can be used for Q1. The coil is wound on an Amidon T68-6 (yellow) toroid core. But, if you have access to other powdered-iron toroids that are 0.68-in OD, with a permeability factor of 8, you may use them (see the Amidon ads in *QST*). Any 3.8-µH coil that can be tapped about ¼ the way from the ground end can be used, also.

The two 100-pF fixed-value capacitors can be silver-mica, polystyrene, or NPO disc-ceramic. If you don't care about frequency stability (drift), you may use ordinary disc-ceramic capacitors. The trimmer capacitor need not be elaborate for this test. One of the Radio Shack plastic 10-pF trimmers will suffice. The tuning control, R1, is not critical. I suggest that you use a 100-kΩ control with a shaft, but a PC-board-mount potentiometer can be used. The value of the control can be anything from 20 kΩ to 100 kΩ for this circuit. The smaller the value of R2, the greater the tuning range. But don't make it less than 22 kΩ under any circumstances. My tuning range was 25 kHz when I used 100 kΩ at R2, but it increased to 55 kHz when R2 was changed to 33 kΩ.

Rather than buy tuning diodes, I chose to use four 2N3904 transistors. They are inexpensive (as little as 10 cents each). You may use only two transistors if you wish, which will compress the tuning range about 10 kHz. The transistors are hooked up so the emitter-base junctions act as tuning diodes. The collectors are not attached to the circuit. You may snip the collector leads or bend them out of the way.

The circuit of Fig. 8 can be tacked together on perforated board. If you want to make a circuit-board type of foundation, glue strips or squares of thin hobby copper to a scrap of Formica®. Or, use a block of wood on which you have mounted multilug terminal strips. Use your imagination: This is the nature of experimentation!

Circuit Testing

Once you're certain all the parts have been wired correctly into your circuit (check two or three times), you are ready to connect the battery and see what happens. Tune your receiver to 40 meters (7.0-7.1 MHz) and connect a short length of wire to the antenna post. Allow this wire to lie near the oscillator (a few inches from L1 of Fig. 7). This will permit ample signal pickup for the receiver.

Set R1 at midrange, then adjust the trimmer (C1) until you hear the signal in your receiver. Next, tune the oscillator by adjusting R1 through its range. As you retune your receiver, you will note that the oscillator frequency has changed in accordance with the setting of R1.

When you first turn on the oscillator, the frequency will drift quickly. It should settle down in 2 or 3 minutes. This is caused by the heating of the Q2 and Q3 junctions when voltage is applied. Also, the junction inside Q1 will change capacitance slightly until it warms up. This condition is known as "short-term drift." If an oscillator continues to drift for long periods (an hour or more), the condition is referred to as "long-term drift."

Later you may want to replace Q2 and Q3 with a 25- or 50-pF air-variable capacitor and repeat the tests. You will find that the drift will practically disappear. With either type of tuning capacitor, you should remember that L1 and the tuning capacitor form a tuned circuit or resonator.

If you can't locate some 2N3904s, use any equivalent NPN transistor. Whatever you select, be sure it has a top-frequency rating (f_T) of at least 50 MHz. A 2N2222 type of transistor can be used at Q2 and Q3, also. The amount of capacitance change versus voltage will vary somewhat with the transistors used.

Good luck, and please do some additional reading on this important subject.

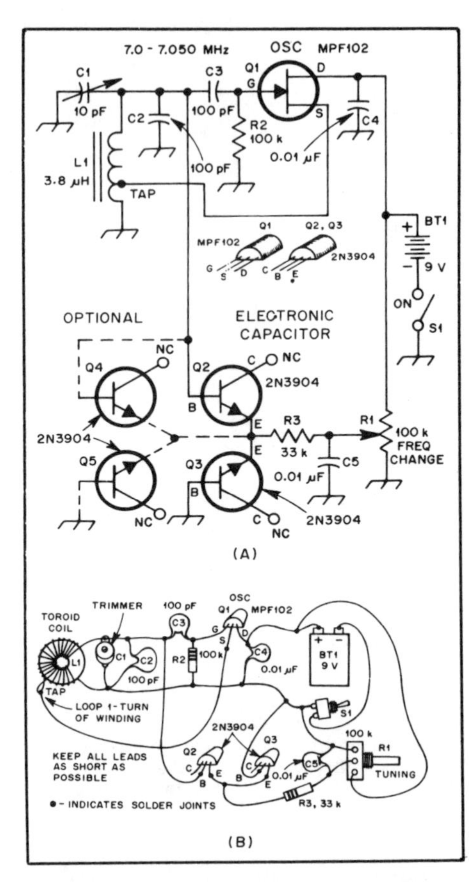

Fig. 8 — The diagram at A is for an oscillator that is tuned by means of diodes (Q2 and Q3), as detailed in the text. The frequency range via R1 is approximately 40 kHz at 40 meters when two transistors are used as diodes. Adding two more diodes (Q4 and Q5 in dashed lines) will increase the tuning range to roughly 50 kHz. Resistors are ¼- or ½-W carbon composition. Fixed-value capacitors are disc-ceramic. A pictorial diagram of the hookup is provided at B.

BT1 — 9-V transistor-radio battery. Radio Shack 23-464 with snap-on connector (Radio Shack 270-325 or equiv.).
C1 — 10-pF maximum capacitance trimmer (Radio Shack 272-1338 or equiv). Not a critical value for this experiment.
C2, C3 — 100-pF NPO capacitor (Radio Shack 272-152 or equiv.).
C4, C5 — 0.01-µF capacitor (Radio Shack 272-131 or equiv.).
Circuit Foundation — Suggest Radio Shack general-purpose type, 276-148.
L1 — 28 turns of no. 24 or 26 enameled wire on an Amidon Assoc. T68-6 (yellow) powdered-iron toroidal core. Amidon Assoc., 12033 Otsego St., North Hollywood, CA 91607 (catalog available). Tap L1 at six turns above grounded end by forming a one-turn loop (twist), then scraping enamel from the loop wire.
Q1 — Junction FET, type MPF102 (Radio Shack 276-2062).
Q2, Q3 — 2N3904 NPN or MPS3904 (Radio Shack 276-2016).
R1 — 100-kΩ audio-taper control with shaft (Radio Shack 271-092). A PC-board-mount thumbwheel control may be used.
R2, R3 — Radio Shack 271-045 and 271-040, respectively.
S1 — SPST slide switch or similar (Radio Shack 275-406).

Understanding FM Transmitters

Part 17: Odds are, you'll operate VHF or UHF FM someday, so why not learn how frequency modulation works.

FM stands for *frequency modulation.* Its cousin is *phase modulation,* or PM. Either method of modulation will permit reception of the transmitted energy by an FM receiver. Here, we'll concentrate on FM and PM transmitters. FM receivers will be addressed in Part 18.

Creating an FM Signal

Two ingredients are necessary to generate an FM radio signal. First, we must have a *carrier* frequency. Second, we need some AF (audio frequency) energy to modulate the carrier. If we allow the audio frequency signal to vary the frequency of the carrier, we'll have an FM signal.

Assume that you're examining the transmitter carrier, as displayed on an oscilloscope (Fig. 1A). Next, suppose a steady audio tone, such as 1 kHz, is generated. It will also appear as a sine wave (Fig. 1B). Note that A and B are on vastly different frequencies, as shown in the illustration. When the 1-kHz audio frequency is applied to the RF carrier, we find a waveform such as that in Fig. 1C.

What is happening here? When the audio energy is applied to the RF carrier, the carrier frequency increases (goes higher in frequency) during half of the audio cycle (positive), and it decreases (shifts lower in frequency) during the negative half of the audio cycle. The RF cycles occupy less time (higher frequency) during the positive period of the modulating cycle, and occupy more time during the negative cycle.

Deviation is the term used for a shift in the RF carrier frequency. Deviation is proportional to the amplitude of our modulating signal; that is, the lower the audio level, the smaller the amount of deviation (frequency swing). Conversely, the higher the audio level, the greater the deviation.

Unlike AM transmitter output, the output from an FM transmitter does not change amplitude during modulation. Rather, the carrier frequency of the FM transmitter swings above and below some center carrier frequency during modulation, but the carrier amplitude remains the same.

Phase Modulation (PM)

The major difference between FM and PM is the method of creating the deviation. Frequency modulation takes place in an oscillator stage. Phase modulation occurs after the oscillator. See Fig. 2.

Another difference is how the frequency of the modulating signal affects the deviation. In FM, the deviation does not change if you change the modulation frequency, assuming the signal level is the same. In PM, on the other hand, the deviation increases with modulating frequency with the signal level held constant.

(A)

WAVELENGTH OF MODULATING SIGNAL

(B)

(C)

Fig. 1 — A graphic representation of FM (and PM). The unmodulated carrier is illustrated at A. The audio-frequency waveform is shown at B. When the modulating energy at B is applied to the RF energy at A, we obtain the display shown at C. (See text).

FM Sidebands

FM signals usually occupy a much wider bandwidth than do AM or SSB signals. Commercial FM stations have peak deviation of 75 kHz, while most amateur and commercial land-mobile FM stations use 5-kHz deviation. These extremes represent wide-band and narrow-band FM, respectively. A peak deviation of 15 kHz was the standard many years ago, but it was abandoned in favor of 5-kHz peak deviation to conserve frequency spectrum in the crowded commercial and amateur bands. In amplitude modulation, there is one set of sidebands, one above, the other below, the carrier frequency. In FM and PM, there can be one, three, five or more sets of sidebands.

The number of sideband pairs that occur during FM or PM operation depends on the ratio between the audio modulating frequency and the carrier-frequency deviation. That ratio is called the *modulation index.* Expressed mathematically:

$$\chi = \frac{D}{m} = \phi \qquad \text{(Eq. 1)}$$

where

χ = modulation index

D = peak deviation (half the difference between the maximum and minimum values of the instantaneous frequency

m = modulation frequency in hertz

ϕ = phase deviation in radians (a

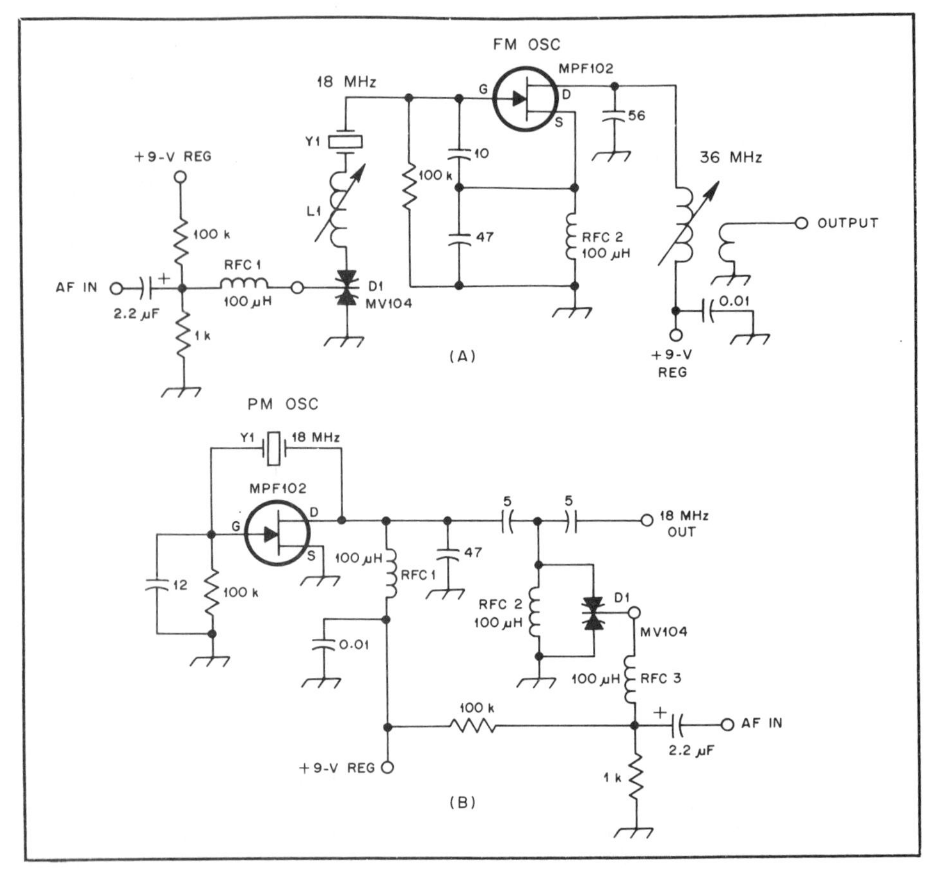

Fig. 2 — Examples of oscillators used for FM and PM generators in a crystal-controlled system. At A, the internal capacitance of D1 changes in accordance with the audio voltage impressed upon it. This change in capacitance causes the crystal frequency to shift above and below the frequency for which it is cut, thereby causing FM. The circuit at B shows how we might generate a PM signal. When audio energy is applied to D1, the phase of the oscillator signal is shifted instantaneously, which results in a frequency shift above and below the frequency of Y1.

radian = $180/\pi$ or approximately 57.3 degrees)

Therefore, if our maximum deviation were 5 kHz (5000 Hz) either side of the center carrier frequency, and the modulating frequency were 1000 Hz, we would obtain the following for the modulation index:

$$\chi = \frac{5000}{1000} = 5 \qquad \text{(Eq. 2)}$$

In the case of PM (with constant amplitude into the modulator), the modulation index is constant, irrespective of modulating frequency. In other words, if a 1-kHz tone causes a 500-Hz deviation, a 2-kHz tone of the same amplitude causes a 1-kHz carrier deviation. In an FM (or PM) system, the ratio of the *maximum* carrier-frequency deviation and the *highest* modulating frequency is called the *deviation ratio.*

The bandwidth of an FM signal depends on the amplitude of the sidebands farthest from the carrier frequency. For a complex waveform such as voice modulation, a good rule of thumb is that the bandwidth is twice the deviation, plus twice the highest modulating audio frequency. Thus, an FM transmitter with 5-kHz deviation modulated by a voice with an upper limit of 3 kHz will have a bandwidth of approximately 16 kHz.

Audio for FM Modulators

To obtain maximum effectiveness from our FM signal, we must ensure that ample audio is available. The average audio level may be increased by means of clipping. This will give the FM signal more apparent volume at the receiver. A simple circuit for creating a clipped and filtered modulating voltage is shown in Fig. 3. Q1 amplifies the audio energy from the microphone. This amplified audio is passed to the speech *clipper* (D1, D2), where the positive and negative peaks of the audio sine wave are squared or clipped. R1 sets the amount of clipping. The clipped audio would cause distortion if it were applied directly to the modulator, so we must filter it first. C1, C2 and R2 of Fig. 3 serve as a simple filter that restores the audio waveform to a sinewave shape.

Some audio power is lost in the filtering process, so we have added Q2 for the purpose of building up the audio level to a sufficient value for modulating the transmitter. The deviation (frequency swing) of the transmitter signal is determined

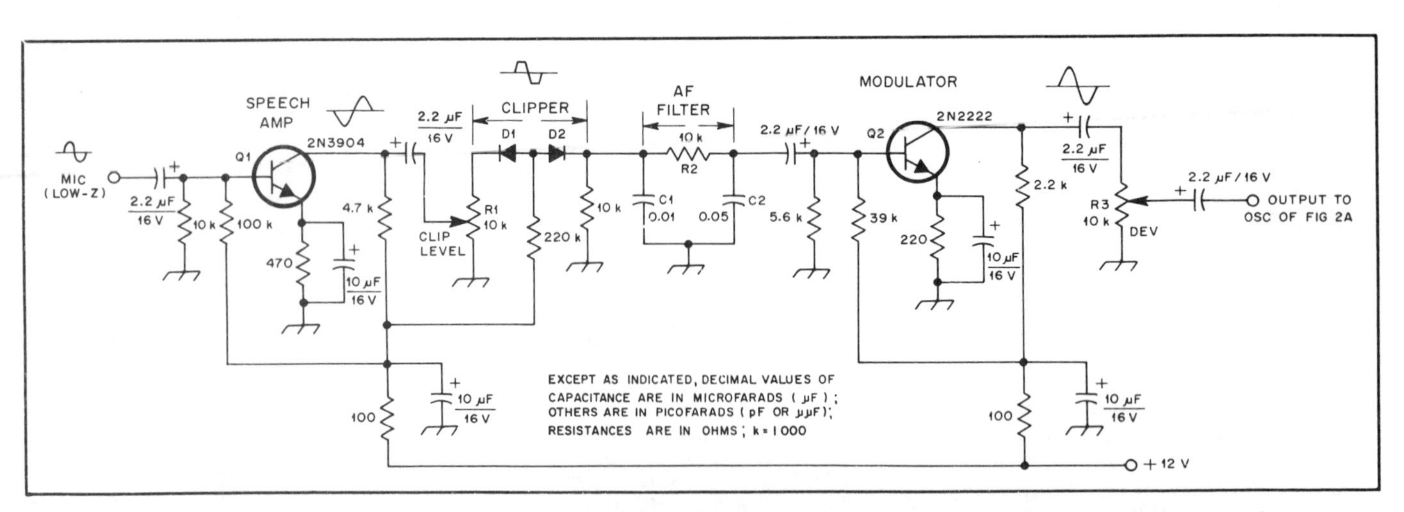

Fig. 3 — Circuit for a simple audio channel that might be used in an FM transmitter. Observe the changes in wave shape as the signal passes through the circuit. Note also the changes in audio signal amplitude. A detailed description of how this circuit operates is given in the text.

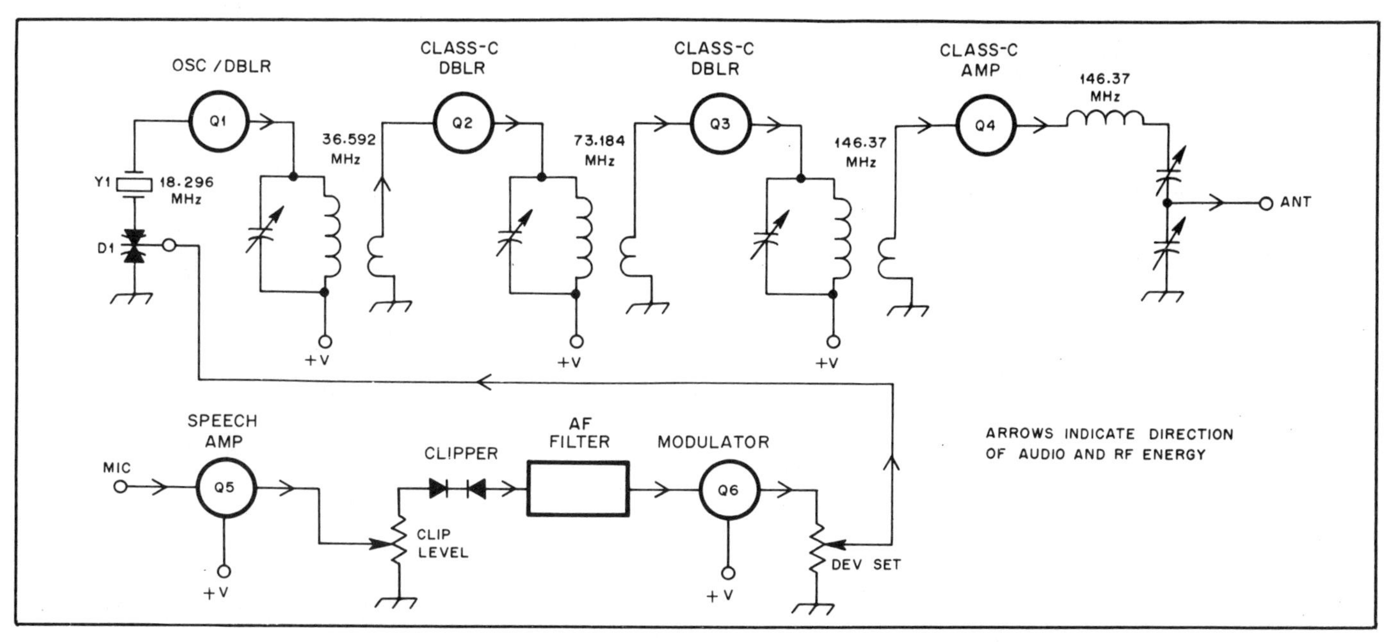

Fig. 4 — Hybrid block diagram of a composite FM transmitter. The frequency of Y1 is multiplied by a factor of eight as the various doubler stages amplify the signal. Similarly, the deviation at Y1 is increased by a factor of eight during the multiplication process. Class-C stages are used throughout the transmitter RF section.

by the setting of R3.

Composite FM Transmitter

How do all the circuits we have discussed fit together? We can consider a typical setup for an amateur FM transmitter, as shown in Fig. 4. This diagram shows the direction of flow (arrows) for the audio and radio frequencies. Consider Q1 the oscillator of Fig. 2A. Q5 and Q6 represent the circuit in Fig. 3. You can see that the oscillator also functions as a frequency doubler. This frequency-doubling action also increases the deviation by a factor of two. The deviation is also doubled in the Q2 and Q3 stages. In this circuit example, the deviation is increased from Q1 to Q4 by a factor of eight. Therefore, in order to have, say, a 5-kHz deviation at 146.37 MHz, we would need only 0.625 kHz of deviation at 18.269 MHz. It is easy to shift the frequency of Y1 that small amount when using D1 as a voltage-variable-capacitor (VVC) diode. The audio energy impressed on D1 causes its internal capacitance to change during the audio cycle, thereby causing the transmitter to swing above and below the carrier frequency.

Use of FM or PM results in perhaps the simplest type of voice transmitter. Very few parts are necessary compared to an SSB transmitter, and we can use class-C transmitter stages without worrying about distortion of the transmitter signal. (Class-A or class-B linear amplifiers are required for SSB transmitters, and their design is somewhat more complicated, to say nothing of the additional components needed.)

Wrap-Up

The aspects of frequency and phase modulation covered in this installment are those you'll be most likely to encounter when taking your amateur license tests. To be fully prepared for exam day, be sure to obtain a copy of the appropriate *ARRL License Manual.* Also, a great deal more about FM circuits and operation can be found in the League publications, *Understanding Amateur Radio* and *FM and Repeaters for the Radio Amateur.*

Glossary

carrier — the RF output from a transmitter, without modulation. It contains no signal information.
clipper — a circuit that limits the peaks of a waveform by clipping or squaring the otherwise rounded positive and negative peaks of a sine wave.
deviation — the amount of frequency swing above and below the FM transmitter carrier frequency when modulating voltage is applied to the low-level RF energy.
FM — frequency modulation.
modulator — a circuit designed to add information to a carrier.
modulation index — pertains to an FM or PM transmitter. The ratio between carrier-frequency deviation (in hertz) and modulating frequency (also in hertz).
PM — phase modulation.
sidebands — bands of frequencies that appear above and below, but close to, the carrier frequency during modulation.
VVC diode — voltage-variable-capacitor diode. The diode internal capacitance changes as the voltage applied to the diode is varied. Sometimes called a *varactor* diode.

Understanding FM Receivers

Part 18: FM receivers aren't much different from AM or CW/SSB receivers. But portions of the circuit are called upon to perform special functions that aren't necessary in other types of receivers.

"Why won't my SSB receiver decipher FM? All I'm getting is gibberish!" Another query could be made: "How come I can't receive CW or SSB on my FM receiver?" The answer is that the method of detecting the various kinds of signals is different. This is necessary because the transmitted signals are processed differently before they are routed to the transmitting antenna. We learned in Part 17 how an FM transmitter creates an FM signal, so you are probably aware that the transmitter output energy is varied above and below the carrier frequency during modulation. This means that a special receiver detector is needed to change the incoming FM signal to comprehensible audio-frequency energy. Generally speaking, the FM receiver circuits ahead of the detector are pretty much the same as those in other types of receivers. That is, we have RF amplifiers, mixers, oscillators and IF amplifiers. The audio chain is the same, also. That much said, let's learn how an FM receiver operates.

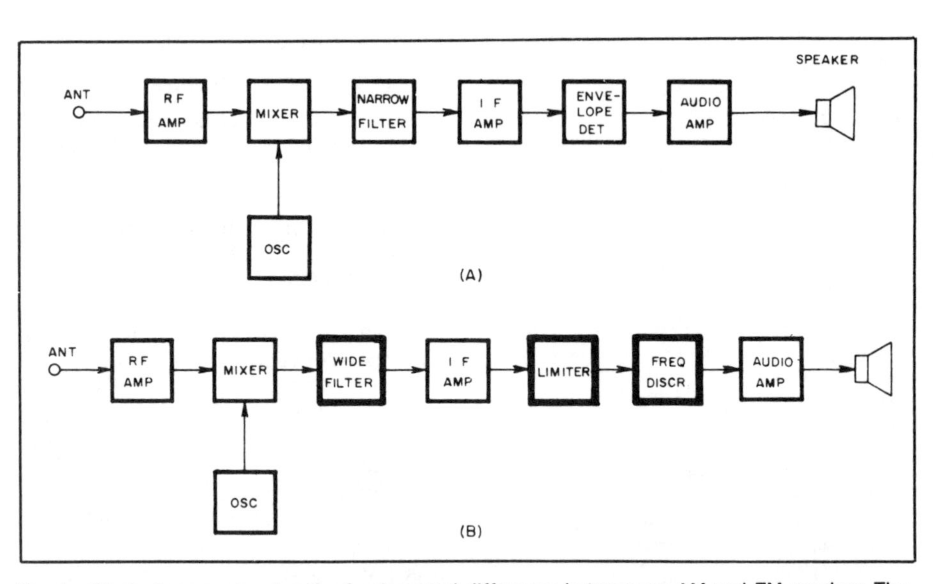

Fig. 1—Block diagram showing the fundamental difference between an AM and FM receiver. The AM version is shown at A; to make it a CW/SSB receiver, change the detector to a product type, then add a BFO that feeds an injection voltage (at the intermediate frequency) to the product detector. Illustration B shows how an FM receiver would be configured.

Comparing Circuits

A block diagram (Fig. 1) illustrates how a CW/SSB receiver compares to an FM radio. The circuits through and including the IF amplifier are identical, except for the effective bandwidth (passband) of the IF filter: A wider filter is needed for FM reception. For example, a 2.4-kHz-wide filter might be used for SSB reception, a 500-Hz filter could be employed for CW work, and a 16-kHz filter might be used in an FM receiver. The filter need only be wide enough to accommodate the bandwidth of the transmitted signal. If the filter has a substantially wider response than the incoming signal bandwidth, unwanted signals (QRM) and noise will be passed along to the detector and audio amplifier.

Both receivers in Fig. 1 are superheterodyne types. A lot of overall receiver gain is needed to ensure high receiver sensitivity. Specifically, an FM receiver needs a gain of more than 1 million to enable us to copy a weak signal that is 1 microvolt (μV) or less at the antenna. I have seen well designed FM receivers that could make an 0.18-μV signal plainly readable above the noise generated *within* the receiver. Most commercial amateur FM receivers are rated at approximately 0.4 μV for what is called "20 dB of quieting," or 20 dBQ. This measurement is made with an audio power meter, calibrated in decibels. The instrument is attached to the receiver output (an 8-ohm resistor replaces the speaker as a dummy load), and audio power is measured across the dummy load. With no signal entering the receiver at the antenna terminal, the audio-gain control is advanced until the audio meter reads, say, 30 dB. Then, a signal generator is fed into the receiver input, and the incoming signal is increased in level until the audio-meter reading drops 20 dB, or to +10 dB on the

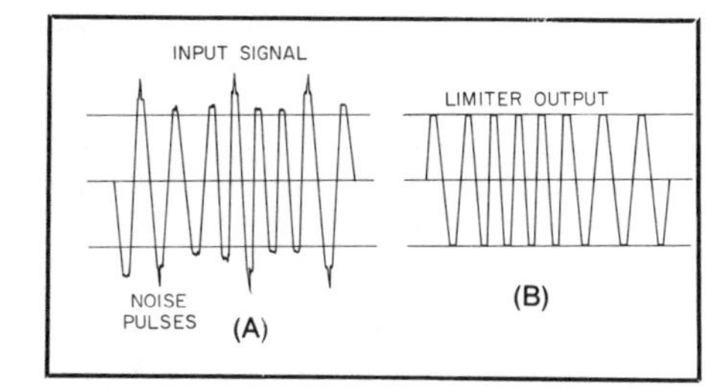

Fig. 2 — Waveforms of an FM signal before passing through a limiter stage (A), and after it has been cleaned up to remove noise and other AM energy by action of the limiter (B). See text.

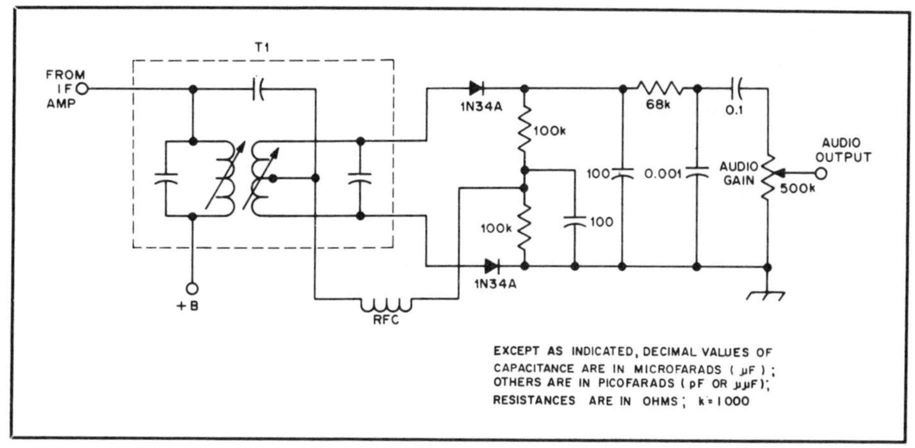

Fig. 3—Typical frequency discriminator circuit that follows a limiter in an FM receiver. This circuit is discussed in the text.

meter scale. The level of signal coming from the signal generator is noted, and that is the signal required for 20 dB of quieting. The lower the level of the input signal, the more sensitive the FM receiver is. A different measuring technique is used with CW/SSB or AM receivers.

Major Circuit Differences

You will notice that in Fig. 1B there is a stage immediately after the IF amplifier that is labeled "limiter." This part of the FM receiver is used to "sanitize" or "launder" the FM signal before it reaches the detector. It saturates (stops providing gain) in the presence of strong signals from the IF amplifier. When this happens, the signal is clipped on both the positive and negative peaks, as would be the case if diodes, reverse-connected, were placed in shunt with the signal path.

Why would we want this to occur? It is vital to take advantage of the limiting feature in order to clip high noise peaks (such as auto-ignition pulses) or any amplitude-modulated energy from other sources. We want only the FM signal to reach the detector. Fig. 2 shows a noisy FM signal (A) entering a limiter, and the cleaned-up signal (B) after leaving the limiter.

A great deal of gain (amplification) is needed ahead of the limiter because it should start functioning as a clipper at 0.2 μV or less. As soon as sufficient signal reaches the limiter, the receiver output (noise) starts quieting. The point on the limiter response curve where limiting action commences is called the "limiting knee." It is at this point that the limiter collector current no longer increases with any buildup in signal amplitude. Modern receivers have ICs rather than individual transistors or tubes in the limiter circuit. An IC may contain several transistor stages; this yields the high gain needed for proper limiter action. If tubes or transistors are used, we might find it necessary to have several such stages in cascade to achieve suitable gain.

FM Detection

There are numerous FM detectors in use today. Among them are the *discriminator, ratio detector, quadrature detector* and *crystal discriminator.* Each has its particular virtues and limitations. The objective in designing an FM detector is to have it respond to FM rather than AM energy. The exact nature of how these detectors operate is rather complex. Detailed information on the subject is contained in *The ARRL Handbook.*

The circuit for a discriminator is given in Fig. 3. The FM signal is changed to AM by means of T1. The T1 secondary *voltage* is 90 degrees out of phase with the *current* in the T1 primary. The signal from the primary winding is routed to the center tap of the secondary winding by means of a coupling capacitor. Next, the secondary voltage combines on each side of the center tap so that the voltage on one side *leads* the primary signal while the other side *lags* by an equivalent amount. When this energy is rectified (changed to dc) by the two diodes of Fig. 3, the two voltages are equal and of opposite polarity. This results in no (zero) output voltage. When voice energy is applied to an FM transmitter, there will be a shift in the received signal frequency, which will lead to a shift in phase at the detector. This phase shift causes an increase in output amplitude on one side of the T1 secondary, along with a corresponding decrease in the other half of the secondary. These differences in the pair of changing voltages (after rectification) create audio output.

Ratio Detector

Fig. 4 illustrates the workings of a ratio detector. You will see some similarity between this circuit and that of Fig. 3. The ratio detector divides the dc voltage into a ratio equal to the ratio of the amplitudes from the two halves of a discriminator transformer secondary winding. The required dc voltage in this circuit is developed across two load resistors, and there is an electrolytic capacitor in shunt across the resistors, as in Fig. 4. The sensitivity of the ratio detector is half that of the discriminator. This is a minor consideration and does not require special attention when the receiver is designed. Ratio detectors are most popular in entertainment FM receivers, whereas discriminators are more common in amateur and commercial land-mobile FM receivers.

Other Considerations

FM receivers do not have automatic gain control (AGC) circuits, but most SSB/CW and AM receivers do. For all practical purposes, the FM limiter acts as an AGC circuit to level the receiver gain after a certain input-signal level is reached. Also, most

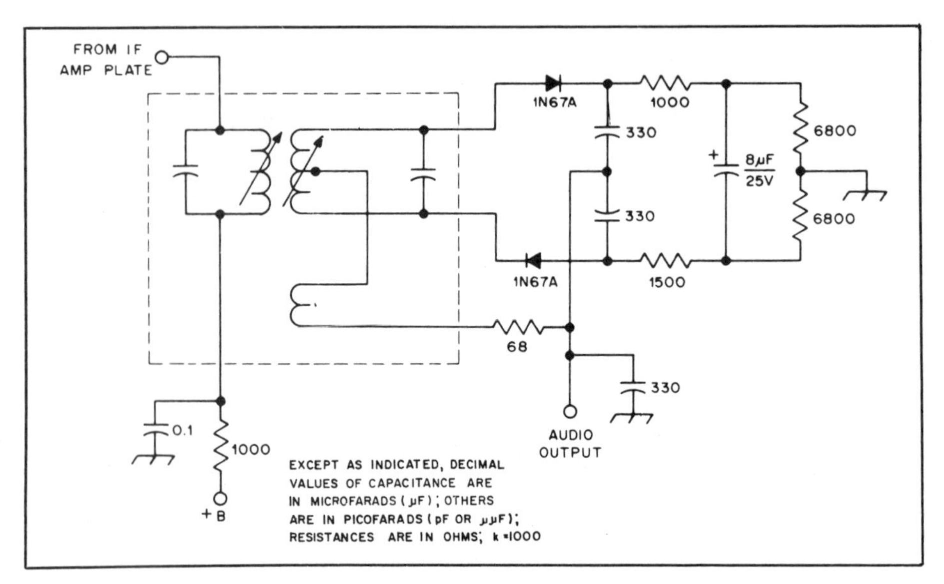

Fig. 4—An FM ratio detector of the type mentioned in the text. It is similar to the discriminator of Fig. 3, but operates in a different manner. Notice that the detector diodes in both circuits are connected in a different polarity arrangement.

amateur FM receivers do not feature continuous-tuning capabilities from the front panel. Rather, a given FM amateur band is covered by means of crystal-controlled frequencies (channels, as some call them) via a frequency-selector switch, or through the use of a synthesizer that tunes in specific frequency increments. Selected frequencies are placed in a memory for instant recall, thereby making it unnecessary to "dial up" a repeater or simplex frequency for day-to-day operation. There is no reason, however, why an amateur FM receiver cannot be made completely tunable for the purpose of covering every kilohertz of a given amateur FM band.

In Summary

We have learned that FM receivers are similar to other types of superheterodyne receivers. The major difference is that FM receivers need a limiter and a special kind of detector. FM now plays a major role in Amateur Radio, so you will certainly become involved with this mode at some point in your amateur career.

Glossary

crystal discriminator — a type of FM detector containing a quartz crystal that replaces the tuned transformer in conventional discriminators.

discriminator — a circuit or device in which amplitude variations are derived from frequency or phase variations. The circuit has many uses, but is popular as a detector in FM receivers.

limiter — in an FM receiver, located immediately ahead of a discriminator or ratio detector to clip AM energy and generally clean up a noisy FM signal.

ratio detector — a type of FM detector that relies on the ratios of voltage and current in the circuit to produce a dc output that can be used as audio-frequency energy.

repeater — a remotely controlled, unmanned (normally) transmitter and receiver that receives signals and retransmits them at high power to extend the effective range of a base station, mobile unit or portable hand-held transceiver.

simplex — as applied to FM operation, direct communications without operating through a repeater. Both stations receive and transmit on the same frequency.

synthesizer — a complex digital circuit that generates precise, very temperature-stable frequencies. It is used as a replacement for conventional VFOs and crystal oscillators.

quadrature detector — an FM detector that depends on the relationship between two periodic functions when the phase difference between them is one-fourth of a period.

Equipping Your First Ham Station

Part 19: You've learned the theory and acquired the needed code speed, and you'll soon be on the air. Now's the time to buy your ham gear and prepare for that first QSO.

New equipment? Used ham gear? Or homemade transmitters and receivers? These are important questions you'll ask yourself when considering the task of putting together an Amateur Radio station. Each of you will want to get maximum benefit from the money you invest, and you will not want to make an error in judgment. Such a mistake in equipment choice could waste money, and the thrill of being a new ham could easily disappear if the gear performed poorly.

In this last part, let's examine the many avenues that are open to us when collecting the necessary items to send and receive communications effectively. The choices open to you are directly related to dollars: There is a definite cost difference between homemade, used and new gear.

Building Your Own Equipment

Some of you are technically skilled. Perhaps you work in the electronics industry and feel comfortable designing or duplicating circuits. In fact, the very reason you have worked for a ham ticket is to be licensed to test transmitters of your own design. That was my motivation when I obtained my license many years ago. You may be more interested in the technology than in operating and ragchewing. If so, you will get many rewards from being an amateur.

But, what about those of you who have no background or special skills in electronics? Building anything other than the simplest of circuits could cost you valuable time and dollars, and you could end up with a unit that works poorly or not at all. For you, I recommend that the building of complex gear be limited at first to kits that are provided by manufacturers with good reputations. Kit building is fun, and it can be educational if you pay attention to the features of the parts and how they work in a circuit.

Assuming that you still want to build some of the station equipment, select uncomplicated things for your first attempts to construct items of convenience or necessity. This list contains a few units that can be easy to build and get working, provided you have a suitable *QST* or *Handbook* presentation to follow.

1) Transmatch (aka antenna tuner or antenna matcher)

2) Field-strength meter for antenna testing

3) Antennas (most popular types)

4) SWR (standing-wave ratio) indicator

5) RF power meter

6) Electronic CW keyer (avoid complex memory keyers)

7) Audio processors for microphones

8) Outboard CW filters (passive or active types)

9) One- or two-transistor transmitters (QRP—low power)

10) Crystal-controlled frequency markers (100, 50 or 25 kHz)

Most of these projects can be completed in an evening or two, and there will be satisfaction connected with the successful building and use of such accessories for your station. As you build more and more simple ham equipment, your skill, knowledge and confidence will increase. This will help you to upgrade your license class. It will also prepare you for some of the more complicated home-constructed circuits.

The Used-Equipment Market

Some of you may prefer to think of used gear as "previously owned equipment," which has a nicer sound. But, no matter what expression you adopt, there are certain dangers lurking in the second-hand-equipment market. Some of the available used gear has "bugs" in it, and that is why the owner decided to get rid of it. This is a chance we must take whenever we purchase second-hand apparatus.

There is money to be saved by avoiding the purchase of new ham gear. Your best opportunity for not being "stung" is to buy the equipment from someone you know and trust. Ask to borrow the item for a day or two while you make up your mind concerning the purchase. Alternatively, you may request a written money-back guarantee for within, say, 10 days of delivery.

My second suggestion is to purchase your used equipment from an established, reputable ham-gear dealer. Be sure there is an option to return the unit if it does not function properly. Several organizations advertise used, reconditioned amateur equipment. Check the ads in *QST*.

The worst-choice plan calls for buying used equipment from classified ads, on-the-air trader nets and trader bulletins. Under these circumstances, you are dealing on a

one-to-one basis with unknown persons; therein lies the gamble. *Caveat emptor* is strictly the rule in this "let the buyer beware" game!

Good Things About Used Gear

Most used equipment operates properly. Rather than spend $1200, for example, when buying a new super transceiver, we may select an 8- or 10-year-old clean (well-cared-for) transceiver for as little as $300. Among the older units that can serve you well as a Novice or Technician are

1) Yaesu FT-101B, EE or E series
2) Drake TR-3 or TR-4 transceivers
3) Kenwood TS-520
4) Ten-Tec Triton 4 and its successor
5) Heath SB-100 or SB-101.

None of these units contains digital frequency readout, but you can do just fine with the analog dials and built-in crystal calibrators. After all, Amateur Radio succeeded marvelously for decades before digital readout was conceived! Certain sophisticated features, such as passband tuning, IF shift, speech processing, memories and outboard redundant VFOs, are missing, but you don't need them to communicate over the airwaves.

Are Transceivers Necessary?

We are the products of a trend toward transceivers that began some years ago with the Collins KWM-series rigs. I must confess that they are a convenience, and provide a more compact station layout than we would have if we chose separate transmitters, receivers and VFOs, as in the old days. But, if you can get a super deal on "separates," don't pass it up for the sake of convenience. A good surplus military receiver, such as the R-390 or 51J1, will work very nicely for you. You may also purchase older civilian receivers like the Collins 75A2, 75A3 or 75A4 for reasonable cost.

For CW transmitters, you may consider a Johnson Viking II, Johnson Valiant, Heath SB-400 (also works on SSB) or one of the old Collins 32V-series AM/CW transmitters. The major inconveniences in using a separate transmitter and receiver is that you will need to (1) employ an antenna changeover relay (controlled by the transmitter), (2) connect a receiver muting line and (3) perhaps use an external VFO for frequency control. Judicious shopping could net you a complete ham station in the 100-W class for as little as $300. The fancy rig can always come later, after you gain experience and learn from other hams the names and model numbers of modern rigs they feel are reliable and cost-effective. You may visit other amateur stations and try the various rigs, thereby developing a first-hand impression of features and performance before laying out money.

Purchasing New Equipment

Today's dealers for new amateur gear attempt to dazzle us with ads that instruct us to "call for prices." Some offer toll-free 800 numbers for this purpose. Personally, I find this annoying, for when I'm considering a new rig I want to compare prices for comparable units of different makes or models. Making several phone calls is time consuming, to say the least. Mail-order purchasing has, however, become a way of life in the USA, and we are almost forced to accept it.

One of the problems relating to mail-order sales is that the dealer you select may be 3000 miles from your location. This makes it difficult and costly to return defective equipment, and new units do come through from time to time in an inoperative condition.

Getting factory service, for foreign-made gear in particular, may be traumatic for you under certain conditions. The quality of the service varies with the manufacturer. It can take weeks to have a warranty repair made, which leaves you high and dry without a rig. Buying a mail-order rig can save money, and the unit may never break down while you own it. But, few of us would consider purchasing a new car from a dealer 2000 miles away! It is an "iffy" proposition, and you should be aware of it. I would definitely check the reputation of the mail-order dealer you decide to become involved with. Those who advertise in *QST* are screened and approved before their ads are accepted, so you're on safe ground with them.

If you have a dealer within driving range of your QTH, I suggest you make an effort to buy from him or her. It's much easier to have problems resolved face-to-face with someone you deal with on a regular basis.

Which Accessories Are Really Necessary?

A new amateur may be told that all manner of additional "goodies" are necessary for his or her ham setup. Knowing which units are essential to routine operating may be difficult for the newcomer. Take, for example, the Transmatch. You may be told that one is needed no matter what type of antenna you are using.

Some hams believe that an SWR reading should be 1:1 at all frequencies. A Transmatch will fool your transmitter into "thinking" an SWR of 1:1 exists, and that is great!

Most antennas exhibit a low SWR over a very narrow range of frequencies within a given amateur band. But no one will know you have an SWR of even 2:1, and many tube-type transmitters can operate effectively at an SWR of 2:1 or more. The shortfall may be, with some solid-state rigs, that the transmitter power will decrease automatically as SWR increases. This is done to prevent the power-amplifier transistors from being destroyed by the effects of high SWR. Your solid-state transmitter may show a power-output drop of only a few watts when the SWR is 2:1, and the difference may not be discernible in the receiver of the other operator. A tube rig will work just fine in the presence of a fairly high SWR.

If you have dipole, vertical or beam antennas that have been adjusted for a low SWR in your favorite parts of the bands, and assuming your feed system uses coaxial cable, you should not need a Transmatch. If, on the other hand, you elect to use a so-called multiband dipole that has tuned, open-wire feeders, you will need a Transmatch and a balun transformer to ensure a low SWR between the transmitter and the feed line: The Transmatch will not correct for a mismatch at the antenna feed point.

If you wish to use electronic keying rather than a straight key or "bug" key, you will want to obtain a keyer or a keyboard keyer. Check the *QST* Ham Ads for a used Curtis, Autek or MFJ keyer. You will also need a CW paddle (key). Beware of keyers that have built-in paddles. Most of them have sloppy mechanical characteristics, and learning to send good CW with those units can be a dreadful challenge. In general, the better the mechanical quality of the paddle, the better your sending. A WW II surplus J-38 straight key is hard to beat for quality in a hand key, as some call them. Whichever key you choose, it should have a heavy base so that it doesn't slip about on the operating table when you are using it.

External audio filters can be very useful for reducing the effects of interference from stations that are nearby in frequency. If your receiver already has a narrow (250-Hz) CW filter in the IF circuit, you may not realize much benefit from a sharp audio filter. But, a great improvement in reception can be had when using a good CW outboard audio filter with older rigs that have no CW filter, or one that is 600 Hz wide. An audio filter will also "lift" weak signals out of the noise to make an otherwise unreadable signal Q5. Your operating preferences (DX or ragchewing) will probably dictate your need for this accessory. These thoughts are applicable to most station accessories. You should evaluate your operating situation versus the style of equipment you have chosen, then decide whether you should invest in additional items. Good circuits for many accessory units are contained in the *ARRL Handbook.*

The Final View

It has been a pleasure to walk with you through these 19 parts of First Steps in Radio. I hope my presentations lead some of you to that first amateur ticket, and that many who are Novices are able to upgrade after studying the basic theory I have covered. Congratulations to all of you, and may your first ham station be a thing of pride and utility.

ABOUT THE

American Radio Relay League

The seed for Amateur Radio was planted in the 1890s, when Guglielmo Marconi began his experiments in wireless telegraphy. Soon he was joined by dozens, then hundreds, of others who were enthusiastic about sending and receiving messages through the air—some with a commercial interest, but others solely out of a love for this new communications medium. The United States government began licensing Amateur Radio operators in 1912.

By 1914, there were thousands of Amateur Radio operators—hams—in the United States. Hiram Percy Maxim, a leading Hartford, Connecticut, inventor and industrialist saw the need for an organization to band together this fledgling group of radio experimenters. In May 1914 he founded the American Radio Relay League (ARRL) to meet that need.

Today ARRL, with more than 160,000 members, is the largest organization of radio amateurs in the United States. The League is a not-for-profit organization that:

- promotes interest in Amateur Radio communications and experimentation
- represents US radio amateurs in legislative matters, and
- maintains fraternalism and a high standard of conduct among Amateur Radio operators.

At League headquarters in the Hartford suburb of Newington, the staff helps serve the needs of members. ARRL is also International Secretariat for the International Amateur Radio Union, which is made up of similar societies in more than 100 countries around the world.

ARRL publishes the monthly journal *QST*, as well as newsletters and many publications covering all aspects of Amateur Radio. Its headquarters station, W1AW, transmits bulletins of interest to radio amateurs and Morse code practice sessions. The League also coordinates an extensive field organization, which includes volunteers who provide technical information for radio amateurs and public-service activities. ARRL also represents US amateurs with the Federal Communications Commission and other government agencies in the US and abroad.

Membership in ARRL means much more than receiving *QST* each month. In addition to the services already described, ARRL offers membership services on a personal level, such as the ARRL Volunteer Examiner Coordinator Program and a QSL bureau.

Full ARRL membership (available only to licensed radio amateurs) gives you a voice in how the affairs of the organization are governed. League policy is set by a Board of Directors (one from each of 15 Divisions). Each year, half of the ARRL Board of Directors stands for election by the full members they represent. The day-to-day operation of ARRL HQ is managed by an Executive Vice President and a Chief Financial Officer.

No matter what aspect of Amateur Radio attracts you, ARRL membership is relevant and important. There would be no Amateur Radio as we know it today were it not for the ARRL. We would be happy to welcome you as a member! (An Amateur Radio license is not required for Associate Membership.) For more information about ARRL and answers to any questions you may have about Amateur Radio, write or call:

ARRL Educational Activities Dept
225 Main Street
Newington CT 06111-1494
(203) 666-1541
Prospective new amateurs call:
800-32-NEW HAM (800-326-3942)

FIRST STEPS
IN RADIO

PROOF OF
PURCHASE